T0257550

Advanced Concepts in Quality Control

Advanced Concepts in Quality Control

Edited by **Theresa Heen**

LANRYE
INTERNATIONAL

New Jersey

Published by Clanrye International,
55 Van Reypen Street,
Jersey City, NJ 07306, USA
www.clanryeinternational.com

Advanced Concepts in Quality Control
Edited by Theresa Heen

© 2015 Clanrye International

International Standard Book Number: 978-1-63240-016-1 (Hardback)

This book contains information obtained from authentic and highly regarded sources. Copyright for all individual chapters remain with the respective authors as indicated. A wide variety of references are listed. Permission and sources are indicated; for detailed attributions, please refer to the permissions page. Reasonable efforts have been made to publish reliable data and information, but the authors, editors and publisher cannot assume any responsibility for the validity of all materials or the consequences of their use.

The publisher's policy is to use permanent paper from mills that operate a sustainable forestry policy. Furthermore, the publisher ensures that the text paper and cover boards used have met acceptable environmental accreditation standards.

Trademark Notice: Registered trademark of products or corporate names are used only for explanation and identification without intent to infringe.

Printed in the United States of America.

Contents

Preface

This book aims to highlight the current researches and provides a platform to further the scope of innovations in this area. This book is a product of the combined efforts of many researchers and scientists, after going through thorough studies and analysis from different parts of the world. The objective of this book is to provide the readers with the latest information of the field.

Rapid development has been made in the past decade in the area of quality control methods and techniques. The objective of this book is to illustrate quality control procedures in several fields, like construction engineering and data quality. A wide range of techniques and methods have been presented in this book covering four major sections: "Quality, General Definitions", "Evaluating Analytical Data", "Quality Control for Biotechnology", and "Planning for Quality Control".

I would like to express my sincere thanks to the authors for their dedicated efforts in the completion of this book. I acknowledge the efforts of the publisher for providing constant support. Lastly, I would like to thank my family for their support in all academic endeavors.

Editor

Part 1

Quality, General Definitions

IA-Quality - General Concepts and Definitions

Ahmed Badr Eldin

Sigma Pharmaceutical Corp.,
Egypt

1. Introduction

The Meanings of "Quality." Of the many meanings of the word "quality," two are of critical importance to managing for quality:
-   "Quality" means those *features of products* which meet customer needs and thereby provide customer satisfaction. In this sense, the meaning of quality is oriented to income. The purpose of such higher quality is to provide greater customer satisfaction and, one hopes, to increase income.

    However, providing more and/or better quality features usually requires an investment and hence usually involves increases in costs. Higher quality in this sense usually "costs more."
-   "Quality" means *freedom from deficiencies* — freedom from errors that require doing work over again (rework) or that result in field failures, customer dissatisfaction, customer claims, and so on. In this sense, the meaning of quality is oriented to costs, and higher quality usually "costs less."

Satisfaction and Dissatisfaction Are Not Opposites. Customer *satisfaction* comes from those features which induce customers to buy the product. *Dissatisfaction* has its origin in deficiencies and is why customers complain. Some products give little or no dissatisfaction; they do what the producer said they would do. Yet they are not salable because some competing product has features that provide greater customer satisfaction. The early automated telephone exchanges employed electromagnetic analog switching methods. Recently, there was a shift to digital switching methods, owing to their superior product features. As a result, analog switching systems, even if absolutely free from product deficiencies, were no longer salable.

Thus Quality can evolve several definitions such as:
-   customer satisfaction and loyalty ;
-   Do right things right.
-   providing a product which is 'fit for the purpose';
-   providing an acceptable product at an acceptable cost;
-   a standard which can be accepted by both the supplier and the customer.
-   the totality of features or characteristics of a product that bear on its ability to satisfy a given need.
-   Fitness for use.

Big Q And Little Q. Definitions of words do not remain static. Sometimes they undergo extensive change. Such a change emerged during the 1980s. It originated in the growing quality crisis and is called the concept of "Big Q."

Table 1 shows how the quality "umbrella" has been broadening dramatically. In turn, this broadening has changed the meanings of some key words. Adoption of Big Q grew during the 1980s, and the trend is probably irreversible. Those most willing to accept the concept of Big Q have been the quality managers and the upper managers. Those most reluctant have been managers in the technological areas and in certain staff functions.

Topic	Content of little Q	Content of big Q
Products	Manufactured goods	All products, goods, and services, whether for sale or not
Processes	Processes directly related to manufacture of goods	All process manufacturing support; business, etc.
Industries	Manufacturing	All industries, manufacturing, service, government, etc., whether for profit or not
Quality is viewed as:	A technological problem	A business problem
Customer	Clients who buy the products	All who are affected, external and internal
How to think about quality	Based on culture of functional departments	Based on the universal trilogy
Quality goals are included:	Among factory goals	In company business plan
Cost of poor quality	Costs associated with deficient manufactured goods	All costs that would disappear if everything were perfect
Evaluation of quality is based mainly on:	Conformance to factory specifications, procedures, standards	Responsiveness to customer needs
Improvement is directed at:	Departmental performance	Company performance
Training in managing for quality is:	Concentrated in the quality department	Companywide
Coordination is by:	The quality manager	A quality council of upper managers

Source: *Planning for Quality,* 2d ed. (1990). Juran Institute, Inc., Wilton, CT, pp. 1–12.

Table 1. Contrast, Big Q and Little Q.

2. Quality: the financial effects

The Effect on Income. Income may consist of sales of an industrial company, taxes collected by a government body, appropriations received by a government agency, tuitions received by a school, and donations received by a charity. Whatever the source, the amount of the income relates in varying degrees to the features of the product produced by the recipient. In many markets, products with superior features are able to secure superior income, whether through higher share of market or through premium prices. Products that are not competitive in features often must be sold at below-market prices.

Product deficiencies also can have an effect on income. The customer who encounters a deficiency may take action of a cost-related nature: file a complaint, return the product, make a claim, or file a lawsuit. The customer also may elect instead (or in addition) to stop buying from the guilty producer, as well as to publicize the deficiency and its source. Such actions by multiple customers can do serious damage to a producer's income.

The Effect on Costs. The cost of poor quality consists of all costs that would disappear if there were no deficiencies — no errors, no rework, no field failures, and so on. This cost of poor quality is shockingly high. In the early 1980s, it was estimated that within the U.S. manufacturing industries, about a third of the work done consisted of redoing what had already been done. Since then, estimates from a sample of service industries suggest that a similar situation prevails in service industries generally.

Deficiencies that occur prior to sale obviously add to producers' costs. Deficiencies that occur after sale add to customers' costs as well as to producers' costs. In addition, they reduce producers' repeat sales.

3. How to manage for quality: the Juran trilogy

To attain quality, it is well to begin by establishing the "vision" for the organization, along with policies and goals. Conversion of goals into results (making quality happen) is then done through managerial processes — sequences of activities that produce the intended results. Managing for quality makes extensive use of three such managerial processes:

1.   1- Quality planning
-     2- Quality control
-     3- Quality improvement

These processes are now known as the "**Juran trilogy**."
A summery for the 3 process is illustrated in table 2.

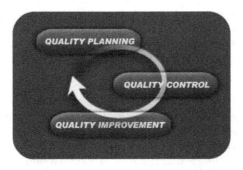

Fig. 1. Juran Triology.

Quality planning	Quality control	Quality improvement
Establish quality goals	Evaluate actual performance	Prove the need
Identify who the customers are	Compare actual performance with quality goals	Establish the infrastructure
Determine the needs of the customers	Act on the difference	Identify the improvement projects
Develop product features that respond to customers' needs		Establish project teams
Develop processes able to produce the product features		Provide the teams with resources, training, and motivation to: Diagnose the causes Stimulate remedies
Establish process controls; transfer the plans to the operating forces		Establish controls to hold the gains

Table 2. The three universal processes of managing for quality. [*Adapted from Juran, J.M. (1989). The Quality Trilogy: A Universal Approach to Managing for Quality. Juran Institute, Inc., Wilton, CT.*]

Inspection and Inspectors. The concepts of inspection and inspectors are of ancient origin. Wall and jewelry paintings in Egyptian tombs show the inspections used during stone construction projects. The measuring instruments included the square, level, and plumb bob for alignment control. Surface flatness of stones was checked by "boning rods" and by threads stretched across the faces of the stone blocks.

Safety and Health of the Citizens. Early forms of protection of safety and health were after-the fact measures. The Code of Hammurabi (c. 2000 B.C.) prescribed the death penalty for any builder of a house that later collapsed and killed the owner. In medieval times, the same fate awaited the baker who inadvertently had mixed rat poison with the flour.

The Industrial Revolution. The Industrial Revolution began in Europe during the mid eighteenth century. Its origin was the simultaneous development of power-driven machinery and sources of mechanical power. It gave birth to factories that soon outperformed the artisans and small shops and made them largely obsolete.

The Twentieth Century and Quality. The twentieth century witnessed the emergence of some massive new forces that required responsive action. These forces included an explosive growth in science and technology, threats to human safety and health and to the environment, the rise of the consumerism movement, and intensified international competition in quality.

An Explosive Growth in Science and Technology. This growth made possible an outpouring of numerous benefits to human societies: longer life spans, superior communication and transport, reduced household drudgery, new forms of education and entertainment, and so on. Huge new industries emerged to translate the new technology into these benefits. Nations that accepted industrialization found it possible to improve their economies and the well-being of their citizenry.

The new technologies required complex designs and precise execution. The empirical methods of earlier centuries were unable to provide appropriate product and process designs, so process yields were low and field failures were high. Companies tried to deal with low yields by adding inspections to separate the good from the bad. They tried to deal with field failures through warranties and customer service. These solutions were costly, and they did not reduce customer dissatisfaction. The need was to prevent defects and field failures from happening in the first place.

Threats to Human Safety and Health and to the Environment. With benefits from technology came uninvited guests. To accept the benefits required changes in lifestyle, which, in turn, made quality of life dependent on continuity of service. However, many products were failure-prone, resulting in many service interruptions. Most of these were minor, but some were serious and even frightening—threats to human safety and health, as well as to the environment.

Thus the critical need became quality.

Expansion of Government Regulation of Quality. Government regulation of quality is of ancient origin. At the outset, it focused mainly on human safety and was conducted "after the fact"—laws provided for punishing those whose poor quality caused death or injury. Over the centuries, there emerged a trend to regulation "before the fact"—to become preventive in nature. This trend was intensified during the twentieth century. In the field of human health, laws were enacted to ensure the quality of food, pharmaceuticals, and medical devices. Licensing of practitioners was expanded. Other laws were enacted relating to product safety, highway safety, occupational safety, consumer protection, and so on.

Growth of government regulation was a response to twentieth-century forces as well as a force in its own right. The rise of technology placed complex and dangerous products in the hands of amateurs—the public. Government regulation then demanded product designs that avoided these dangers.

To the companies, this intervention then became a force to be reckoned with.

4. The rise of the consumerism movement

4.1 How to think about quality

Consumers lacked expertise in technology. Their senses were unable to judge which of the competing products to buy, and the claims of competing companies often were contradictory. When products failed in service, consumers were frustrated by vague warranties and poor service.

"The system" seemed unable to provide recourse when things failed. Individual consumers were unable to fight the system, but collectively they were numerous and hence potentially powerful, both economically and politically. During the twentieth century, a "consumerism" movement emerged to make this potential a reality and to help consumers deal more effectively with these problems. This same movement also was successful in stimulating new government legislation for consumer protection.

Intensified International Competition in Quality. Cities and countries have competed for centuries. The oldest form of such competition was probably in military weaponry. This competition then intensified during the twentieth century under the pressures of two world wars. It led to the development of new and terrible weapons of mass destruction. A further stimulus to competition came from the rise of multinational companies. Large companies had found that foreign trade barriers were obstacles to export of their products. To get around these barriers, many set up foreign subsidiaries that then became their bases for competing in foreign markets, including competition in quality. The most spectacular twentieth-century demonstration of the power of competition in quality came from the Japanese. Following World War II, Japanese companies discovered that the West was unwilling to buy their products—Japan had acquired a reputation for making and exporting shoddy goods. The inability to sell became an alarm signal and a stimulus for launching the Japanese quality revolution during the 1950s. Within a few decades, that revolution propelled Japan into a position of world leadership in quality. This quality leadership in turn enabled Japan to become an economic superpower. It was a phenomenon without precedent in industrial history.

5. Quality to center stage

The cumulative effect of these massive forces has been to "move quality to center stage." Such a massive move logically should have stimulated a corresponding response—a revolution in managing for quality. However, it was difficult for companies to recognize the need for such a revolution—they lacked the necessary alarm signals. Technological measures of quality did exist on the shop floors, but managerial measures of quality did not exist in the boardrooms. Thus, except for Japan, the needed quality revolution did not start until very late in the twentieth century. To make this revolution effective throughout the world, economies will require many decades—the entire twenty-first century. Thus, while the twentieth century has been the "century of productivity," the twenty-first century will be known as the "century of quality." The failure of the West to respond promptly to the need for a revolution in quality led to a widespread crisis. The 1980s then witnessed quality initiatives being taken by large numbers of companies.

Most of these initiatives fell far short of their goals. However, a few were stunningly successful and produced the lessons learned and role models that will serve as guides for the West in the decades ahead.

Lessons Learned. Companies that were successful in their quality initiatives made use of numerous strategies. Analysis shows that despite differences among the companies, there was much commonality—a lengthy list of strategies was common to most of the successful companies. These common strategies included:

Customer focus: Providing customer satisfaction became the chief operating goal.

Quality has top priority: This was written into corporate policies.

Strategic quality planning: The business plan was opened up to include planning for quality.

6. IB-quality control - general concept

Quality Control Defined. "Quality control" is a universal managerial process for conducting operations so as to provide stability—to prevent adverse change and to "maintain the status quo."

To maintain stability, the quality control process evaluates actual performance, compares actual performance to goals, and takes action on the difference.

Quality control is one of the three basic managerial processes through which quality can be managed. The others are quality planning and quality improvement,. The Juran trilogy diagram (Figure 2) shows the interrelation of these processes. Figure 2 is used also to describe the relationships between quality planning, quality improvement, and quality

control and the fundamental managerial processes in total quality management. What is important for this section is to concentrate on the two "zones of control." In Figure 2 we can easily see that although the process is in control in the middle of the chart, we are running the process at an unacceptable level of waste. What is necessary here is not more control but improvement—actions to change the level of performance. After the improvements have been made, a new level of performance has been achieved. Now it is important to establish new controls at this level to prevent the performance level from deteriorating to the previous level or even worse. This is indicated by the second zone of control. The term "control of quality" emerged early in the twentieth century (Radford 1917, 1922). The concept was to broaden the approach to achieving quality, from the then-prevailing after-the-fact inspection, to what we now call "defect prevention." For a few decades, the word "control" had a broad meaning which included the concept of quality planning. Then came events which narrowed the meaning of "quality control." The "statistical quality control" movement gave the impression that quality control consisted of using statistical methods. The "reliability" movement claimed that quality control applied only to quality at the time of test but not during service life. In the United States, the term "quality control" now often has the narrow meaning defined previously.
The term "total quality management" (TQM) is now used as the all-embracing term.

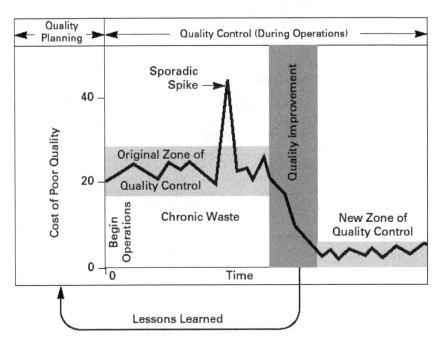

Fig. 2. The Juran trilogy diagram. (*Juran Institute, Inc., Wilton, CT.*)

In Europe, the term "quality control" is also acquiring a narrower meaning. Recently, the European umbrella quality organization changed its name from European Organization for Quality Control to European Organization for Quality. In Japan, the term "quality control" retains a broad meaning.

Their "total quality control" is roughly equivalent to our term "total quality management." In 1997 the Union of Japanese Scientists and Engineers (JUSE) adopted the term total quality management (TQM) to replace total quality control (TQC) to more closely align themselves with the more common terminology used in the rest of the world. The quality control process is one of the steps in the overall quality planning sequence,. Figure 3 shows the input-output features of this step. In Figure 3 the input is operating process features developed to produce the product features required to meet customer needs. The output consists of a system of product and process controls which can provide stability to the operating process.

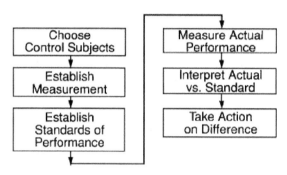

Fig. 3. The input-output diagram for the quality control process.

7. The relation to quality assurance

Quality control and quality assurance have much in common. Each evaluates performance. Each compares performance to goals. Each acts on the difference.

However they also differ from each other. Quality control has as its primary purpose to maintain control. Performance is evaluated during operations, and performance is compared to goals during operations. The resulting information is received and used by the operating forces. Quality assurance's main purpose is to verify that control is being maintained. Performance is evaluated after operations, and the resulting information is provided to both the operating forces and others who have a need to know. Others may include plant, functional, or senior management; corporate staffs; regulatory bodies; customers; and the general public.

The Feedback Loop. Quality control takes place by use of the feedback loop. A generic form of the feedback loop is shown in Figure 4. The progression of steps in Figure 4 is as follows:

1. A sensor is "plugged in" to evaluate the actual quality of the control subject—the product or process feature in question. The performance of a process may be determined directly by evaluation of the process feature, or indirectly by evaluation of the product feature—the product "tells" on the process.
2. The sensor reports the performance to an umpire.
3. The umpire also receives information on what is the quality goal or standard.
4. The umpire compares actual performance to standard. If the difference is too great, the umpire energizes an actuator.
5. The actuator stimulates the process (whether human or technological) to change the performance so as to bring quality into line with the quality goal.

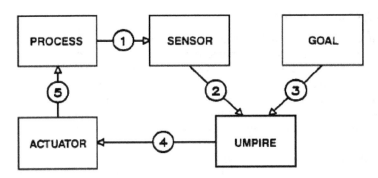

Fig. 4. The generic feedback loop. (*Making Quality Happen, Juran Institute, Inc., senior executive workshop, p. F-3, Wilton, CT.*)

8. The elements of the feedback loop

The feedback loop is a universal. It is fundamental to any problem in quality control. It applies to all types of operations, whether in service industries or manufacturing industries, whether for profit or not. It applies to all levels in the hierarchy, from the chief executive officer to the work force, inclusive. However, there is wide variation in the nature of the elements of the feedback loop. In Figure 5 a simple flowchart is shown describing the quality control process with the simple universal feedback loop imbedded.

The Process. In all of the preceding discussion we have assumed a process. This may also be human or technological or both. It is the means for producing the product features, each of which is a control subject. All work is done by a process which consists of an input, labor, technology, procedures, energy, materials, and output. For a more complete discussion of process.

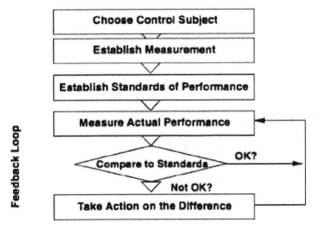

Fig. 5. The qualty control process. (*"Quality Control", Leadershitp for the Quality Century, Juran Institute, Inc., senior executive workshop, p. 2, Wilton, CT.*)

9. Deming chain reaction

The so-called Deming Chain Reaction was actually borrowed from a model that Walter Shewhart developed. He probably borrowed the idea from another thinker. Basically the idea was for management to move away from thinking about quality as a desirable outcome, to thinking about quality as a competitive strategy. Competitive strategy as a concept has been around for centuries. A person selling an item similar to that sold by another can compete on price, by selling it for less money. Perhaps the seller may try to compete by adding extras, gift-wrapping, for example. Technical companies compete by being technology leaders and being on the cutting edge of new developments. There are no end to methods to compete. But some methods are more effective in the long run than others. It is not a mistake that Deming's first published book on the subject was entitled "On Quality, Productivity and Competitive Position". In the book, he sets forth the reasons why emphasis on quality leads to productivity improvement and how that is a very effective competitive strategy in the long run. Phil Crosby in the early 80s in his book, "Quality is Free" pointed out that improving quality lowered cost. But Deming had shown this to the Japanese 30 years earlier. And, Deming pointed out the benefits of developing a competitive strategy based on quality. One of the problems in talking about quality is that many people have pre-conceived notions of quality is. For some it is meeting specifications. Joseph Juran defines it as 'meeting customer requirements'.

Fig. 6.

Zero Defects was Crosby's nostrum, but is really just another way of saying quality is meeting specfications. Deming's ideas are much broader than that and are, perhaps, best captured with the phrase 'continual improvement'. This term connotes the ongoing nature of the strategy. According to Deming, quality is not a state to be achieved in manufacturing, but is, rather, an ongoing company-wide effort at continual improvement. What Bill Conway called "the process – the way everyone thinks, talks, works and acts every day." After all the nonsense is stripped away, the fact is that Japanese automakers (Toyota, Honda and Nissan) make better cars than American automakers (GM, Ford and Chrysler) and have done now for years. Buyers are not idiots. They understand value, and realize that better quality at the same or lower cost is excellent value. End of story. But the implications of the story are not just confined to the auto industry. Cameras, computers, appliances, power tools, earthmoving equipment, and more have fallen from America's basic manufacturing industries to a legacy of plant closings, job losses and dwindling revenues and profits.

10. Quality control: what is new?

Recent decades have witnessed a growing trend to improve the effectiveness of quality control by formal adoption of modern concepts, methodologies, and tools. These have included: Systematic planning for quality control, with extensive participation by the operating personnel Formal application of the feedback loop, and establishment of clear responsibility for the associated
decisions and actions Delegation of decisions to the work force through self-control and self-inspection Wide application of statistical process control and the associated training of the operating personnel A structured information network to provide a factual basis for decision making A systematic process for corrective action in the event of sporadic adverse change Formal company manuals for quality control, with periodic audits to ensure up-to-dateness and conformance.

11. Quality control in pharmaceutical industries

11.1 Responsibilities of quality control unit

a. There shall be a quality control unit that shall have the responsibility and authority to approve or reject all components, drug product containers, closures, in-process materials, packaging material, labeling, and drug products, and the authority to review production records to assure that no errors have occurred or, if errors have occurred, that they have been fully investigated. The quality control unit shall be responsible for approving or rejecting drug products manufactured, processed, packed, or held under contract by another company.
b. Adequate laboratory facilities for the testing and approval (or rejection) of components, drug product containers, closures, packaging materials, in-process materials, and drug products shall be available to the quality control unit.
c. The quality control unit shall have the responsibility for approving or rejecting all procedures or specifications impacting on the identity, strength, quality, and purity of the drug product.
d. The responsibilities and procedures applicable to the quality control unit shall be in writing; such written procedures shall be followed.

12. References

Phillips-Donaldson, Debbie, American Society for Quality 37 (5): 25–39(2000).

Bunkley, Nick, Joseph Juran, Pioneer in Quality Control, Dies", New York Times, 37,March 3 (2008).

Dr. Joseph M. Juran JURAN'S QUALITY HANDBOOK McGraw-Hill, p. 79, (2004).

Parasuraman, A., Zeithami, Valarie A., and Berry, Leonard L, Journal of Marketing, Fall, pp. 41–50 (1985).

Koura, Kozo Societas Qualitas, Japanese Union of Scientists and Engineers, Tokyo, p. 180 - 186 (1991).

Dawes, Edgar W., and Siff, Walter, ASQC Annual Quality Congress Transactions, pp. 810–816(1993).

2

GLP: Good Laboratory Practice

Isin Akyar

Acibadem University Faculty of Medicine Department of Medical Microbiology
Turkey

1. Introduction

In the early 70's FDA (United States Food and Drug administration) have realized cases of poor laboratory practice throughout the United States. FDA decided to check over 40 toxicology labs in-depth. They revealed lot dishonest activities and a lot of poor lab practices. Examples of some of these poor lab practices found were equipment not been calibrated to standard form, therefore giving wrong measurements, incorrect or inaccurate accounts of the actual lab study and incompetent test systems. Although the term "good laboratory practice" might have been used informal already for some time in many laboratories around the world GLP originated in the United States and it had a powerfull effect world wide.

2. History of Good Laboratory Practice (GLP)

GLP is an official regulation that was created by the FDA in 1978. The OECD (Organisation for Economic Co-operation and Development) Principles of Good Laboratory Practice were first created by an Expert Group on GLP set up in 1978 under the Special Programme on the Control of Chemicals. The GLP regulations that are accepted as international standards for non-clinical laboratory studies published by the US Food and Drug Administration in 1976 supplied the basis for the work of the Expert Group, which was guided by the United States and consisted experts from the following countries and organisations: Australia, Austria, Belgium, Canada, Denmark, France, the Federal Republic of Germany, Greece, Italy, Japan, the Netherlands, New Zealand, Norway, Sweden,Switzerland, the United Kingdom, the United States, the Commission of the European Communities, the World Health Organisation and the International Organisation for Standardisation. Eventually after United States other countries started making GLP regulations in their home countries. (Lori et al., 2009)

2.1 Those Principles of GLP were officially suggested for use in member countries by the OECD Council in 1981. They were set about as an essential part of the Council Decision on Mutual Acceptance of Data in the Assessment of Chemicals, which expresses that"data denoted in the testing of chemicals in an OECD member country in accordance with OECD Test Guidelines and OECD Principles of Good Laboratory Practice shall be accepted in other member countries for the aims of assessment and other uses relating to the protection of man and the environment".

2.1.1 The work of the OECD associated with chemical safety is fulfilled in the Environmental Health and Safety Division. The Environmental Health and Safety Division publishes free-off charge documents in six different series: Testing and Assessment; Principles on Good Laboratory Practice and Compliance Monitoring; Pesticides; Risk Management; Chemical Accidents and Harmonization of Regulatory Oversight in Biotechnology.

2.1.2 In spite of the fact that there are many national guidelines setting Good Laboratory Practice, the one guideline that is most universally accepted by the various national guidelines is the regulation of GLP through the Principles of Good Laboratory Practice of the Organisation of Economic Cooperation and Development (OECD), since these have been discussed by an international panel of experts and have been agreed on at an international level; they also form the basis for the OECD Council Decision/Recommendation on the Mutual Acceptance of Data in the Assessment of Chemicals which has to be regarded as one of the cornerstone agreements amongst the OECD member states with regard to trade in chemicals and to the removal of non-tariff barriers to trade. Besides the utilisation of the OECD Guidelines for the Testing of Chemicals, they restated the application of GLP Principles and the establishment of consorted national GLP compliance monitoring programmes as necessary parts of the mutual acceptability of data. The working group of experts who had createded the OECD Principles of Good Laboratory Practice also proceeded to inform and publish guidance for the Monitoring Authorities with regard to the introduction of procedures essential for the monitoring of industry's compliance with these Principles, as well as guidance with respect to the actual conduct of the necessary control activities such as laboratory inspections and study audits. (OECD, 1998).

2.1.3 Thus, the Principles of Good Laboratory Practice (GLP) have been developed to promote the quality and validity of test data used for determining the safety of chemicals and chemical products. Its principles are postulated to be followed by test facilities carrying out studies to be referred to national authorities for the purposes of assessment of chemicals and other uses in regards with the protection of man and the environment. Good laboratory practice might be used to detect collusion, but it could also serve to protect the researcher from unfounded allegations. In this manner, the application of the basic rules of GLP could be benefit even to a instution or laboratory.

2.1.4 Definition of GLP

The quality is the capability to systematically produce the same product to meet the same specifications time after time. GLP was altered to protect the integrity and quality of laboratory data used to back up a product application. The definition of the term "Good Laboratory Practice" itself, which identifies GLP as "a quality system related with the organisational process and the conditions under which non-clinical health and environmental safety studies are planned, performed, monitored, recorded, archived and reported." can be considered as an example of a brief and accurate definition. GLP describes good practices for non-clinical lab studies that support research or marketing approvals for FDA-regulated products(Seiler, 2005).

2.1.5 Purpose of GLP

Everyone makes mistakes that's why GLP is needed. GLP principles are a good idea even if you are not required to follow the standards. There are some simple rules such as: Say What You Do (with written standard operating procedures), do what you say (follow the procedures), be able to prove it (with good record keeping) (Jean Cobb, 2007).

2.1.6 The principles of good laboratory practice (GLP) is to support the development of quality and validity of test data used for determining the safety of chemicals and chemicals product (Clasby, 2005).

Hence GLP aims to decrease the occurrence of mistakes or mix-ups through large and specific labelling requirements. The registered information can be provided by demonstrating the application of the correct item in the stated amounts to the pertinent test systems.

2.1.7 GLP experience is important to employers in some cases. An employer may find it useful if you have: Practical experience with working on a study according to the GLP principles.

Good planning is the greater half of success. With a perfect propose in mind and a well figured out and defined testing procedure is it achiavable to acquire an evaluable outcome of a study. GLP places a high degree of reliance upon creating and following a pre-defined study plan.

2.2 The principles of good laboratory practice

Good Laboratory Practice is based on four principles:

The Management; The Quality Assurance; The Study Director; and The National Compliance Monitoring Authority. All of them serve important functions in the concordancy of performing and monitoring safety studies, and it should be kept in mind that all of them are required for GLP to achieve quality data.

2.2.1 Although GLP differs from other quality systems in aspects that are important not only for the traceability of data but especially for the full reconstructability of the study, there are certain co-occurances between GLP and other quality systems like accreditation schemes. (Seiler, 2005).

2.2.2 The aim of this chapter will be to give enough information about the GLP in details with the test facility organisation and personel, the facilities of quality assurance programme, test system, archive and waste disposal, apparatus, material, and reagents, physical, chemical, biological test systems, receipt, handling, sampling and storage and characterisation of the test and reference items, standard operating procedures, performance of the study, reporting of study results, storage and retention of records and materials.

2.2.3 The concerns of the chapter may be summarized as follows:

1. Test facility management
2. Quality assurance programme
3. Meeting the requirements of the test facility
4. Equipment
5. Receipt, handling, sampling and storage
6. Standard operating procedures.
7. Performance of the study.
8. Reporting of study results
9. Storage and retention of records and materials.

3. Test facility management

Test facility means the persons, premises and operational units that are necessary for conducting the non-clinical health and environmental safety study.

3.1 The term "test facility" may include several "test sites", at one or more geographical locations, where phases or components of a single overall study are conducted and does not only include buildings, rooms and other premises, but that it includes also the people who are working there and are liable for performing these studies (Seiler, 2005). For multi-site studies the test facility considers the site at which the Study Director is located and all individual test sites, which individually or collectively can be considered to be test facilities. The test facility should be of appropriate size, construction and location to meet the requirements of the study. It should be designed safe enough to get the validation results confidently. Research laboratories where test/reference item characterisation considering determination of identity, purity/strength, stability, and other related activities is conducted, one or more agricultural or other in- or outdoor sites where the test or control item is applied to the test system are the different test sites in the test facility. And in some cases, a processing facility where collected commodities are treated to prepare other items where collected specimens are analysed for chemical or biological residues, or are otherwise evaluated. (OECD, 1998).

3.1.1 Properties of biological test systems are generally more complex and mutable than the ones of physical/chemical test systems. Hence biological test systems need very careful characterisation in order to guarantee the quality and integrity of the data derived from them. The outcome of a study may be influenced by the state and condition of the test system at the time of the study which has special importance with regard to the reconstructability. The GLP Principles, in uttering the requirements for the accomodation and siting of these systems, for their maintenance and utilization, and for the associating documentation, aims at supplying the essential basis for confidence into the results obtained from biological test systems. A test item should only be used in studies if it can safely be regarded as being in its pure, unspoilt and not decomposed. Any change in the properties of the test item may lead to spurious and erroneous results, and to wrong interpretations of the effects the test item is supposed to have produced. Stability testing will lead to the definition of a time interval within which the test item will stay in this state, and as a result "expiry" or "re-analysis" dates have to be mentioned on the label of the test item container. With this necessity GLP aims to reduce the possibility that an item will be used in a study which does no longer correspond to the item that had been intended for testing. The aim of any safety testing is to analyze possible effects of the test item on the test system. Therefore, the effects observed in any test system should be traceable to the application of the item which was the designated subject of the study.

3.1.2 After the conduct of the respective safety test, in order to find out this even retrospectively, the documentation on the test item has to fulfil a number of requirements:

3.1.3 There must be documented proof that one item that had been intended to be tested indeed reached the sensitive parts of the test system confirming that the effects observed had really been originated by the test item, and that the application of this item to man or the environment would therefore not be expected to result in any effects other than those which can be concluded from the observed ones in the test systems utilised. "Tidiness" is a crucial point with consideration to the general claims on the test facility. When the laboratory bench is filled up with clean and dirty instruments, glassware some of which are being used and some are not, it is not so easy to locate all the materials needed for a specific activity.

3.1.4 Tidiness therefore has both functions of inspiring trust into the quality of the work performed, and facilitate the performance of the daily activities according to the quality standards. Tidiness makes the life easier to survive a compliance monitoring inspection, if even under the stress the technician can find the folder with the SOPs at once like without trying to find a treasure.

3.1.5 A test facility needs a Management, a Study Director, a Quality Assurance Unit, study personnel and a person responsible for the archives (Seiler, 2005).

3.1.6 Test Facility Management should guarantee that these Principles of GLP are requested in its test facility. General Requirements for GLP consists of appropriately qualified personnel, adequate resources , appropriate procedures for: sanitation, health precautions, clothing, test protocol development, test methods, data analysis, report development, appropriately qualified study director, quality assurance function. Test site management should be aware of the fact that the test facility management may be liable to inspection by the national GLP compliance monitoring authority of the country in which the test site is located.

3.1.7 "The Study Director" has overall responsibility for the technical conduct of the study, as well as for the interpretation, analysis, documentation, and reporting of results, and represents the single point of study control." (OECD, 1998).

3.1.8 The GLP Principles are designed to avoid the factors that would endanger the reconstructability of a study, by giving the only and final responsibility for the GLP compliant conduct of a study to one single person, the Study Director. For each nonclinical laboratory study, a scientist or other professional of appropriate education, training, and experience should be identified as the study director.

3.1.9 The Study Director has to be aware of all possible circumstances that might affect the quality and integrity of a study. There should be communication between the Study Director and other personnel including all scientists involved in study conduct, in order to be kept at the forefront of developments in a study, and to be able to act, as considered appropriate, on unforeseen developments. All information has to be passed to the Study Director. He should make or at least acknowledge all the decisions. In such special circumstances where the Study Director cannot exercise his immediate control, the responsibilities of a Study Director may be extended to other individuals such as specialised scientists (Seiler, 2005).

3.1.10 When the Study Director cannot exercise immediate supervision, at each test site study procedures may be controlled by a member of the staff, called the Principal Investigator. The Principal Investigator means an individual responsible for the conduct of certain defined phases of the study, acting for the Study Director. The Study Director has the final responsibility of for the overall quality and integrity of the study. He cannot share this responsibility with any other individual involved in the study. Nonetheless, the Principal Investigator should take the responsibility for the defined, delegated part of the study, he is not responsible for the study plan, and he can not approve any improvements to it. The general management must have a stiff interpretation and working agreement with the test site management as to how and by whom the Quality Assurance Programme (QAP) will be carried out.(OECD, 1998).

3.1.11 Approved original and revised Standard Operating Procedures should be used in the studies. There should be a Quality Assurance Programme with assigned personel for each study an individual with the proper qualifications, training, and experience is designated by the management as the Study Director before the study is initiated. Personnel should clearly understand the functions that they are going to carry out, training should be provided when needed. Standard Operating Procedures should be established and followed. They should be appropriate and technically valid.

3.1.12 The GLP Compliance Statement signed by the Study Director in the final study report is the declaration that gives the Regulatory Authority the guarantee for a appropriately performed, valid study. The results and conclusions of the study can be trusted to reflect the real data obtained in the study (Seiler, 2005).

4. Quality assurance programme

Quality control is the process, procedures and authority used to accept or reject all components, drug product containers, closures, in-process materials, packaging material, labeling and drug products and the authority to review production records to assure that no errors have occurred, that they have been fully investigated. The quality and reliability of test data count on the state and condition of the test system which is used in its production.

This is meant to be the control of a number of technical features and specifications which are needed to ensure the integrity of the system and the quality of the data generated. In a study for compliance with GLP, the most important aspects may be characterised as "suitability", "capacity" and "integrity" (OECD, 1998).

4.1 "Trust is Good, Control is Better" says an old proverb. The quality which is supposed to be achieved in GLP is not a quality which can be controlled by easy, numerical or other means, but it is the şcontrol over the intrinsic quality of a test facility and its studies. Only through this independence a reliable assurance of the studies inherent quality that can be achieved. (Seiler, 2005).

4.1.1 The test facility should have a documented Quality Assurance Programme to guarantee that studies performed comply with these Principles of Good Laboratory Practice. The Quality Assurance Programme should be performed by an individual or by individuals designated by. These staff should be familiar with the test procedures and directly responsible to management.This individual(s) should not be involved in the conduct of the study being assured(OECD, 1998). It must be clear that what the exact area of responsibility is for the defined individual, what exactly is to be done at those test sites where such "phases" are conducted in delegating parts or"phases" of a study through the terms of appointment for the Contributing Scientist or the Principal Investigator (Seiler, 2005).

4.1.2 As the person responsible for the overall conduct of the study, to the Study Director's management, and to the latter's Quality Assurance Programme, there should be a full, frank flow of information to the responsible test site management, to the responsible Principal Investigator(s) and to the Study Director. In the same way, for notification of critical activities it should be essential to assure effective communications from the Study Director and/or Principal Investigators to the quality assurance personel. Because of the complex nature of field studies, and the fact that the exact time of certain activities will depend upon local weather or other conditions flexible quality assurance procedures may be required. The geographical spread of test sites may mean that quality assurance personnel will also need to manage langquage differences in order to communicate with local study personnel, the Study Director, Principal Investigators and test site management. Independent from the test sites, the written reports of quality assurance personnel must reach both management and the Study Director.Those reports receipt by management and the Study Director should be documented in the raw data.

4.1.3 The Quality Assurance personnel should be responsible of maintaining copies of all approved study plans and Standard Operating Procedures in use in the test facility and have access to an up-to-date copy of the master Schedule, verifying that the study plan contains the information required for compliance with these Principles of Good Laboratory Practice, conducting inspections to determine if all studies are conducted in accordance with these Principles of Good Laboratory Practice. Inspections should also determine that study plans and Standard Operating Procedures have been made available to study personnel and are being followed. The study plan allows Quality Assurance: to monitor compliance of the study plan with GLP; to assess the clarity and consistency of the study plan; to identify the critical phases of the study; and to plan a monitoring programme in relation to the study (OECD, 1998).

4.1.4 Study plans and Standard Operating Procedures should be determined by the inspections and they should have been available to study personnel and are being followed. In the final reports it should be confirmed that the methods, procedures, and observations are accurately and completely described, and that the reported results accurately and completely reflect the raw data of the studies

4.1.5 Inspection of facilities and experimental activities is one of the tools of Quality Assurance for ascertaining and guaranteeing the continued obeyence to the rules of GLP in a test facility inside the studies performed. Since it is recognised that randomly conducted inspections will be sufficient to ensure compliance with, the GLP Principles do not necessitate a fixed supervision. These inspections should involve those parts of a study that have particular importance for the validity of the data and the conclusions to be drawn therefrom, or where deviations from the rules of GLP would most heavily have a powerfull effect on the integrity of the study. Quality Assurance thus has to find a balance in their inspectional activities, evaluating the study type and "critical phases", in order to achieve a well supported view of the GLP compliance at the test facility and within the studies conducted. It is clear that any deviations from the rules of GLP that are observed in these inspections should be corrected. The audit of the final report, hence serves to ascertain the quality and integrity of the specific study with its detailed assessment of GLP compliance throughout the study and with its concomitant review of all relevant information, records and data. It is the responsibility of management to provide policies, guidelines, or procedural descriptions to ensure that this statement reflects Quality Assurance's acceptance of the Study Director's GLP compliance statement. The Quality Assurance statement has two functions: Serving to demonstrate that Quality Assurance has adequately monitored the conduct and progress of the study, from the first check of the study plan for GLP conformity to the audit of the final report as a "second opinion" on the completeness of the reporting and the adequacy of raw data coverage and providing the study with the seal of approval by attesting to the GLP compliant conduct. Thus, the Quality Assurance statement has a particular importance for the assessment of the study's integrity and validity. The Quality Assurance statement should show that the study report accurately reflects the study's raw data.

4.1.6 Before signing the Quality Assurance statement, Quality Assurance should ensure that all issues raised in the Quality Assurance audit, i.e. in the audit report to the Study Director and to management, have been addressed through appropriate changes of the final report, that all agreed actions have been completed, and that no additional changes have been made to the report which would require a further report audit. Through management policy it should certainly be made clear that the Quality Assurance statement would only be completed if the Study Director's claim to GLP compliance can be supported(Seiler, 2005).

Laboratories use various supplied materials in studies conducted in compliance with the GLP Principles. Suppliers have attempted to produce products which satisfy users' obligations as set out in the GLP Principles.

4.1.7 Accreditation can be especially useful to suppliers. Often accreditation schemes monitor members' implementation of national and international standards thus, a supplier or manufacturer's accreditation certificate may signify to the customer the satisfactory implementation of a standard in addition to other aspects of accreditation.

4.1.8 It is recommended that suppliers seek membership, where feasible and/or appropriate, in national accreditation schemes. Although accreditation is a useful complementary tool to support compliance with the GLP Principles, it is not an acceptable alternative to GLP compliance nor will it lead to international recognition in the context of meeting the requirements for the mutual acceptance of data as set out in the OECD Council Acts. (OECD, 1998).

As an example ISO 17025 and GLP comparison can be considered (Table 1).

ISO Members	OECD Members
The same standard for all ISO	Different regulations in different countries
Designed for repetitive studies	Designed for single studies
Description of Quality System in Quality Manual	Description of Quality System in SOPs
General statements for responsibilities of personnel	Very specific responsibilities of personnel
No specific requirements for storage of records and reports	Specific requirements for storage, retention and archiving
No study plans required (standardized methods should be used)	Study plan required for each study
Written operating procedures without specific format	SOPs with detailed requirements for format and content
Analysis methods must be verified through inter-laboratory test (Proficiency testing)	Validation through inter-laboratory tests not required
Documented complaints procedures	In case of problems, only course of law
Storage of test samples and data until client accepts results	Storage of test samples according to local regulatory requirements

(Fox , 2011)

Table 1. ISO 17025 and GLP comparison.

5. Meeting the requirements of the test facility

The GLP principles do not address the question of the specific requirements for the location of an archive, except that it should be "of suitable size, construction and location to meet requirements". Therefore there is complete freedom for every test facility to define the location of its archives and to designate the proper locations for each type of materials to be stored (Seiler, 2005). Before they can be considered as GLP compliant General Requirements Facilities need to conform to a number of general rules. The facilities should be designed for the best suitability to the studies that are to be performed within. Some comfort for the employees comes of course with all the requirement of study quality, which means that the people working in a facility should certainly have sufficient room to move around in order to be able to perform the duties which the study calls for, and to perform them in a manner compatible with the quality, integrity and validity of the study. This is acknowledged absolutely in the general requirement that a test facility should be of appropriate size, construction and location, for both meeting the requirements of the study and minimising disturbance that would interfere with the validity of the study. jürgThe test facility should have a sufficient and suitable number of rooms or areas to assure the isolation of test systems and the isolation of individual projects, involving substances or organisms known to be or suspected of being biohazardous. There should be storage rooms or areas as needed for supplies and equipment and should provide adequate protection against infestation, contamination, and/or deterioration. Facilities for handling test and reference items should be planned. To prevent contamination or mix-ups, there should be separate rooms or areas for receipt and storage of the test and reference items, and mixing of the test items with a vehicle(Figure 1).

Fig. 1. There should be separate working areas in the laboratory.

5.1 Handling and disposal of wastes should be carried out in such a way as not to risk the integrity of studies. This includes provision for appropriate collection, storage and disposal facilities, and decontamination and transportation procedures. This policy is to assure that reagents used are specified in the standard operating procedure. Purchasing and testing should be handled by a quality assurance program. Reagents and solutions should be labeled, deteriorated or outdated reagents and solutions should not be used. The opening date should be recorded. They should be stored under ambient temperature and the expiration date should be considered(Lori et al , 2009).The equipments should be appropriately designed, adequate throughput capacity, appropriately located and routinely maintained & calibrated (Clasby, 2005).

5.1.1 In order to guarantee the quality of the data, appropriate conditions should be established and maintained for the storage, housing, handling and care of biological test systems. At the experimental starting date of a study, test systems should be free of any disease or condition that might interfere with the purpose or conduct of the study. If necessary to maintain the integrity of the study, test systems that become diseased or injured during the course of a study should be isolated and treated. Any diagnosis and treatment of any disease before or during a study should be recorded. Records of source, date of arrival, and arrival condition of test systems should be maintained. Biological test systems should be acclimatised to the test environment for an adequate period before the first administration/application of the test or reference item. All information needed to properly identify the test systems should appear on their housing or containers. Individual test systems that are to be removed from their housing or containers during the conduct of the study should bear appropriate identification, wherever possible. During use, housing or containers for test systems should be cleaned and sanitised at appropriate intervals. Any material that comes into contact with the test system should be free of contaminants at levels that would interfere with the study. Bedding for animals should be changed as required by sound husbandry practice. Use of pest control agents should be documented. Test systems used in field studies should be located so as to avoid interference in the study from spray drift and from past usage of pesticides (OECD, 1998).

5.1.2 The important principles can be summarised as follows:

There should be a unique identification for the study and all of its parts. All original observations in a study should be at once clearly and legibly recorded. The recording should be permanent and corrections should be made so as not to obscure the original entry; for all corrections the respective reasons have to be provided. All records should be in the form of bound notebooks or on continuously numbered sheets. All entries and corrections to them should be dated and initialled. Records related to the test system itself should be gathered and preserved. Specimens should be clearly identified so as to allow full traceability. At the end of a study, all raw data should be assembled, catalogued and archived. Archiving should support for secure storage of all raw data, samples and specimens, together with any other documents such as study plan and study report. (Jürg P. Seiler, 2005).

6. Equipment

6.1 Equipment, including validated computerised systems, used for the generation, storage and recovery of data, and for controlling environmental factors relevant to the study should be suitably located and of appropriate design and adequate capacity. Equipment records should include: name of the equipment and manufacturer, model or type for identification, serial number, date equipment was received in the laboratory, copy of manufacturers operating instruction(s). Equipment used in a study should be periodically inspected, cleaned, maintained, and calibrated according to Standard Operating Procedures. Records of these activities should be maintained. Calibration should be traceable to national or international standards of measurement. Instrumentation validation is a process necessary for any analytical laboratory. Data produced by "faulty" instruments may give the appearance of valid data. The frequency for calibration, re-validation and testing depends on the instrument and extent of its use in the laboratory. Chemicals, reagents, and solutions should be labelled to indicate identity, expiry date and specific storage instructions. Information concerning source, preparation date and stability should be available. The expiry date may be extended on the basis of documented evaluation or analysis. If a mistake is made, original data should not be obscured. Instead of this, a single strikeout should be drawn and a reason code should be added, later the date should be changed. Whenever an instrument's performance is outside the "control limits" reports must be discontinued

(Cobb, 2007). Equipment and materials used in a study should not interfere adversely with the test systems. (OECD, 1998).

6.1.1 Equipment used for the generation of physical/chemical data should be suitably located and of proper design and adequate capacity. The integrity of the physical/chemical test systems should be ensured. Appropriate conditions should be established and maintained for the storage, housing, handling and care of biological test systems, in order to ensure the quality of the data. Standardization, calibration, and verification are the definitions which have particular importance for the equipments. The difference between those should be well understood and performed by the laboratory personnel: Verification is the external check of equipment accuracy. It is the check balance accuracy against weights at laboratory. There is no adjustment.

6.1.2 In calibration equipment is adjusted based on comparison to certified or known reference materials. The balance is adjusted after comparison to certified weights by trained professional. Standardization is made by comparison with similar equipments, such as using two thermometers of similar design to compare readings.

6.1.3 While monitorizing the study laboratory staff should always have the following questions on mind: Was the equipment functioning properly? Who performed the work, what was the date, and what specific parameters did they use? Was there a problem? How was the problem fixed? Were there any problems with the reagents and solutions?

Fig. 2. Laboratory equipment should routinely be maintained and calibrated.

6.1.4 The GLP Principles do not suggest or require any specific time intervals for such activities. Cleaning and maintenance intervals may be different from one type of equipment to the other, and such intervals may as well depend on the frequency of use or the workload imposed on the respective equipment. On the other hand the question of the correct frequency of such activities should be considered as a scientific one, calling for the expert judgement of the responsible scientists.

6.1.5 Generally the manufacturer's manuals provide useful signs or suggestions for cleaning and maintenance intervals. These same aspects are valid also for calibration frequencies, where in some cases calibration is routinely performed before each measurement, while in other cases the respective frequencies may be set in an arbitrary manner. The key point in the consideration of maintenance and calibration frequencies is the necessary assurance of data validity. In some cases it would be necessary to ensure the traceability of the calibrations performed to "national or international standards of measurement". The results of a study can be relied on only as far as the study itself is being appropriately conducted. Suitability of apparatus, materials and reagents is thus one of the key points in this judgement. Computerised systems have taken over an ever increasing part of different tasks in various areas within our daily lives. They are used during the planning, conduct and reporting of studies for a variety of purposes, including the direct or indirect data capture from automated instruments, the recording, processing, reporting, general management and storage of data, as well as in controlling and steering functions in numerous kinds of equipment. For these different activities, computerised systems can be of varying complexity from a simple, microchip controlled instrument up to a complex laboratory information management system (LIMS) with multiple functions. Whatever the scale of computer involvement, the GLP Principles have to be followed. The correct application of the GLP Principles to ensure compliance of computerised systems with the GLP rules may, however, pose some problems, which might be regarded to stem at least in part from the very origins of GLP. All computerised systems used for the generation, measurement or assessment of data intended for regulatory submission should be developed, validated, operated and maintained in ways which are compliant with the GLP Principles. Appropriate controls for security and system integrity must also be adequately addressed during the whole life cycle of any computerised system(Seiler, 2005).

6.1.6 All equipment used in a GLP context have to satisfy the specified requirements of the users. For computerised systems the evidence of suitability is provided by the validation procedure. This has to start with the exact definition of the user requirements which have subsequently to be translated into proof of adequate operation of the system in the actual environment. With this prospective validation assurance it should be provided that the computerised system will perform the tasks designed to execute in a correct, reproducible and reconstructable way.

6.1.7 Computerised systems associated with the conduct of studies bound for regulatory submission should be of appropriate design, adequate capacity and suitable for their intended purposes. There should be appropriate procedures to control and maintain these systems, and the systems should be developed, validated and operated in a way which is in compliance with the GLP Principles. The demonstration that a computerised system is suitable for its intended purpose is of fundamental importance and is referred to as computer validation. The validation process provides a high degree of assurance that a computerised system meets its pre-determined specifications. Validation should be undertaken by means of a formal validation plan and performed prior to operational use. (OECD, 1998).

6.1.8 Whether any system has been fully and prospectively validated or has just been retrospectively evaluated and qualified, there is a need for continued maintenance of the validation status to be sure of the continuence of data validity. This is accomplished through formal procedures that require any changes to the system to be fully documented. Data integrity will, however, not only depend on the validation status of the system, but also, and to a very important extent, on the security measures developed for the utilisation of the system. Through the requirement of documented security procedures for the protection of hardware, software and data from corruption, unauthorised modification, or loss, GLP intends to provide for continuous data integrity. In general terms, security issues can be divided into measures of physical security, i.e. measures that can be instituted on the facility and apparatus level, and logical security, i.e. those that are related to software security at the access level (Seiler, 2005).

6.1.9 Physical location of computer hardware, peripheral components, communications equipment and electronic storage media should be considered. Extremes of temperature and humidity, dust, electromagnetic interference and proximity to high voltage cables should be avoided unless the equipment is specifically designed to operate under such conditions. Consideration must also be given to the electrical supply for computer equipment and, where appropriate, back-up or uninterruptable supplies for computerised systems, whose sudden failure would affect the results of a study. Adequate facilities should be provided for the secure retention of electronic storage media. (OECD, 1998).

6.1.10 Because of various reasons, in every test facility there may be computerised systems which have not been formally validated. Their use in a GLP environment should still be required, clear proof of their suitability can only be obtained through an evaluation of their past and actual performance. In order to get reconstructability and transparency, this proof has to be planned and documented, resulting in a final conclusion on the past, present and future suitability of the respective system. In this way GLP aims at providing evidence for the correct functioning of the computerised system and for estimating the extent of GLP compliance.

7. Receipt, handling, sampling and storage

Sampl e tracking vary among laboratories. Receipt, handling, sampling and storage should be prepared appropriately. Records including test item and reference item characterisation, date of receipt, expiry date, quantities received and used in studies should be maintained. Handling, sampling, and storage procedures should be identified in order that the homogeneity and stability are assured to the degree possible and contamination or mixup are precluded (Seiler, 2005). They should maintain the unmistakable connection between a set of analytical data and the samples from which they were obtained. Original source of samples must be recorded and unmistakably connected with the set of analytical data (Cobb, 2007). Records including test item and reference item characterisation, date of receipt, expiry date, quantities received and used in studies should be maintained. Handling, sampling, and storage procedures should be identified in order that the homogeneity and stability are assured to the degree possible and contamination or mix-up are precluded. Storage container(s) should carry identification information, expiry date, and specific storage instructions.

7.1 Receipt and storage areas for specimens must be separate from storage areas for pesticide formulations and other test or reference items. Areas used for specimen and sample preparation, instrumentation, calibration of sprays, reference Standard preparation, and for washing glassware should be adequately isolated from each other and from other functions of the laboratory which might introduce contamination. Storage areas for test and reference items at all test sites should be environmentally monitored, if required, to assure conformance with established stability limits for these materials. Test and reference items should not be placed in the same storage containers with collected test system specimens and other materials of low concentrations which are being stored for shipment to the analytical laboratory or to off-site archives. There should be adequate storage and disposal facilities available for pesticide and related wastes such that there is no potential for cross-contamination of test systems, of test or reference items or of collected specimens. (OECD, 1998). Storage container(s) should carry identification information, expiry date, and specific storage instructions Each test and reference item should be properly identified. For each study, the identity, including batch number, purity, composition, concentrations, or other characteristics to appropriately define each batch of the test or reference items should be known. In cases where the test item is supplied by the sponsor, there should be a mechanism, developed in co-operation between the sponsor and the test facility, to verify the identity of the test item subject to the study. The stability of test and reference items under storage and test conditions should be known for all studies. If the test item is administered or applied in a vehicle, the homogeneity, concentration and stability of the test item in that vehicle should be determined. A sample for analytical purposes from each batch of test item should be retained for all studies except short-term studies. A well thought-out concept of logistics is needed for receiving, storing, handling and disposing test items, together with provisions for the adequate documentation of all procedures connected with test item handling. One aspect in this area of test item logistics is the physical location of these activities, and the GLP Principles underline the importance of identifying adequate facilities for them.

Fig. 3. Laboratory records of receipt, handling and storing should be carefully maintained.

7.1.1 While receipt and storage involves mainly the handling of closed containers, the opening of such a container exposes the test item to the facility environment and leads consequently to the possibility of contamination of either the test item or the environment. Moreover, the greater the number of different test items to be performed, the greater the danger that somebody would. Therefore, work in the special area where test items are prepared for application has to be carefully organised. For weighing of the test item and its mixing with the vehicle, it should be made compulsory that only one test item would be present in that area at any one time. Special attention has to be given to such areas where test, control and reference items are prepared for in vitro studies.

7.1.2 In such studies, the term "contamination" does not only mean "contamination by traces of other items" but also contamination by microorganisms, etc.,hence necessitating areas where the preparation of these items for the application in the study could be performed under aseptic conditions. By the same reason, GLP mandates that the available test item storage locations should be separate from the rooms or areas containing test systems in order to prevent excessive exposure of the systems to test items other than the intended one.

7.1.3 Of course, the storage facilities should supply adequate conditions to save the identity, purity and stability of the test items. Therefore it is necessary that storage areas at different temperature levels, for storage at room temperature or in refrigerators and deep freezers. Also protection from light, humidity or oxygen may be necessary for special cases. Also, there are security aspects to be mentioned. A suitable limitation for access to the test items should be advisable. It is very important that a good, accurate accounting system should be in place, which could be used to reconstruct the course of test item utilisation. (Seiler, 2005).

8. Standard Operating Procedures (SOP)

According to EPA(Environmental Protection Agency) GLP regulations, "Raw data" means any laboratory worksheets, records, memoranda, notes, or exact copies thereof, that are the result of original observations and activities of a study and are necessary for the reconstruction and evaluation of the report of that study. Logbooks for recording temperatures or equipment use, repair, and maintenance, field or laboratory notebooks, forms for field or laboratory observations, training reports, computer printouts, recorded data from automated instrument are examples of raw data. It's so hard and not necessary for anyone remember all these details and that's one of the functions of the Standard Operating Procedures (SOPs).

8.1 In FDA it is said that :"If it is not documented..., it did not happen!" or, it's a rumor!" GLPs SOPs Can't do Guarantee "good science", guarantee good documentation, replace common sense, prevent all mistakes (Cobb, 2007). SOPs are written procedures for a laboratories program. They are approved protocols indicating test objectives and methods. Standard Operating Procedures are intended to ensure the quality and integrity of the data generated by the test facility. Revisions to Standard Operating Procedures should be approved by test facility management (OECD, 1998).

8.1.1 They define how to carry out protocol-specified activities. SOPs are most often written in a chronological listing of action steps. They are written to explain how the procedures are supposed to work SOP of routine inspection, cleaning, maintenance, testing and calibration, actions to be taken in response to equipment failure, analytical methods, definition of raw data, keeping records, reporting, storage, mixing, and recovery of data. (Standard Operating Procedures should have been written and approved by test facility management that are intended to ensure the quality and integrity of the data generated by that test facility. Revisions to Standard Operating Procedures should be approved by test facility management. Each separate test facility unit or area should have at once available current Standard Operating Procedures relevant to the activities being performed therein. Published text books, analytical methods, articles and manuals may be used as supplements to these Standard Operating Procedures. Deviations from Standard Operating Procedures related to the study should be documented and should be acknowledged by the Study Director and the Principal Investigator(s). SOPs are written, approved procedures that describe routine activities that are specific for daily operations at each facility. SOPs should allow appropriately qualified personnel to perform a procedure once trained.

8.1.2 The details given under each heading are to be considered as illustrative examples. Room preparation and environmental room conditions for the test system, procedures for receipt, transfer, proper placement, characterisation, identification and care of the test system, test system preparation, observations and examinations, before, during and at the conclusion of the study, handling of test system individuals found in a severe position or dead during the study, collection, identification and handling of specimens ,siting and placement of test systems in test conspiracy should be reviewed. And also operation of Quality Assurance personnel in planning, scheduling, performing, documenting and reporting inspections should be examined. Personnel should perform the same tasks using the same procedures. SOPs should accurately reflect how routine tasks are performed written by each facility based on their specific field and/or laboratory operations. Laboratory management must be sure that the SOPs used in the laboratory are useful in daily operations. They should be scientifically sound. And they should always be updated as necessary, rewrites should be the part of the routine process. While writing SOP guidelines there must be some precautions such as avoiding restrictive language such as "vortex for exactly 1 minute" but include clear instructions such as "vortex until homogenized" if that satisfies the purpose. Unnecessary steps should not be added such as "consult the manual" unless personnel are required to follow this step (Cobb, 2007). Study personnel should easily access to the study plan and appropriate Standard Operating Procedures should be applicable to their involvement in the study. It is their responsibility to comply with the instructions given in these documents. Study personnel should exercise health precautions to minimise risk to themselves and to ensure the integrity of the study. Standard Operating Procedures (SOPs) are intended to describe procedures that are routinely employed in the performance of test facility operations. Indeed they are defined as "documented procedures which describe how to perform tests or activities normally not specified in detail in study plans or test guidelines." The definition moreover implies that SOPs should describe all steps in the performance of an activity in such a detailed way that somebody not familiar with this activity might be able to perform it correctly and without having to recourse to outside help (Seiler, 2005).

8.1.3 It is suggested that test site personnel should follow test site SOPs. When they are required to follow other procedures specified by the Study Director, for example SOPs provided by the test facility management, this necessity should be identified in the study plan (OECD, 1998).

9. Performance of the study

Performance of the Study should be monitorized carefully. All the standards supplied by the GLP should be followed from the beginning of the study to the end by the final report. For each study, a written plan should exist prior to the initiation of the study (Seiler, 2005). The study plan should contain the following information: Identification of the study, the test item and reference item, information concerning the sponsor and the test facility, dates, test methods, issues (where applicable)and records. (OECD, 1998).

9.1 The study plan should be approved by dated signature of the Study Director and verified for GLP compliance. Deviations from the study plan should be described, explained, recognized and dated in a timely fashion by the Study Director and/or Principal Investigator(s) and maintained with the study raw data.

9.1.1 In the study plan the identification of the study, the test item and reference item information should exist: A descriptive title; a statement which reveals the nature and purpose of the study; Identification of the test item by code or name; The reference item to be used. Information Concerning the Sponsor and the Test Facility should be declared. It should comprise: Name and address of the sponsor, any test facilities and test sites involved, Study Director, Principal Investigator(s), and the phase(s) of the study delegated by the Study Director and under the responsibility of the Principal Investigator(s) with the date of approval of the study plan by signature of the Study Director, of the study plan by signature of the test facility management and sponsor if required by national regulation or legislation in the country where the study is being performed, the proposed experimental starting and completion dates, reference to the OECD Test Guideline or other test guideline or method to be used, the justification for selection of the test system characterisation of the test system, such as the species, strain, substrain, source of supply, number, body weight range, sex, age and other pertinent information. It should also contain the method of administration and the reason for its choice; The dose levels and/or concentration(s), frequency, and duration of administration/application; detailed information on the experimental design, including a description of the chronological procedure of the study, all methods, materials and conditions, type and frequency of analysis, measurements, observations and examinations to be performed, and statistical methods to be used. Specimens from the study should be identified to confirm their origin. Such identification should enable traceability, as appropriate for the specimen and study. The study should be conducted in accordance with the study plan. All data generated during the conduct of the study should be recorded directly, punctually, correctly, and legibly by the individual entering the data. These entries should be signed or initialled and dated. Any change in the raw data should be made in order to understand the previous entry easily, should indicate the reason for change and should be dated and signed or initialled by the individual making the change.

9.1.2 Computerised system design should always supply for the retention of full audit trails to show all changes to the data without obscuring the original data. It should be possible to associate all changes to data with the persons having made those changes. Reason for changes should be given.

10. Reporting of study results

All studies generate raw data that are the original data gathered during the conduct of a procedure. They are essential for the reconstruction of studies and contribute to the traceability of the events of a study. Raw data are the results of the experiment upon which the conclusions of the study will be based. Some of the raw data may be used directly, and some of them will be treated statistically. The results and their interpretations provided by the scientist in the study report must be a true and accurate reflection of the raw data.

10.1 A final report should be prepared for each study. The study report, like all the other scientific aspects of the study, is the responsibility of the Study Director. He/she must ensure that it describes the study accurately. Reports of Principal Investigators or scientists involved in the study should be signed and dated by them. The final report should be signed and dated by the Study Director to indicate acceptance of responsibility for the validity of the data.If necessary, corrections and additions to a final report should be in the form of amendments. Amendments should clearly specify the reason for the corrections or additions and should be signed and dated by the Study Director. The Study Director is responsible for the scientific interpretation included in the study report and is also responsible for declaring to what extent the study was conducted in compliance with the GLP Principles. The GLP Principles list the essential elements to be included in a final study report.

10.1.1 The final report should include, the following information: A descriptive title; identification of the test item by code or name, characterisation of the test item including purity, stability and homogeneity. Information concerning the sponsor and the test facility should imply; name and address of the sponsor, any test facilities and test sites involved, the study Director, the Principal Investigator(s) and the phase(s) of the study, delegated and scientists having contributed reports to the final report, experimental starting and completion dates.

10.1.2 A Quality Assurance Programme statement listing the types of inspections made and their dates, including the phase(s) inspected, and the dates any inspection results should be reported to management and to the Study Director and Principal Investigator(s). This statement should also serve to confirm that the final report reflects the raw data. It should contain the Description of Materials and Test Methods. A summary of results should be given. All information and data required by the study plan; A presentation of the results, including calculations and determinations of statistical significance; An evaluation and discussion of the results and, where appropriate, conclusions. It should imply the location(s) where the study plan, samples of test and reference items, specimens, raw data and the final report are to be stored.

10.1.3 A computerised system to be used in a GLP area should include both the dating and timing of the original entry and the retention of a full audit trail. Such identification could be possible either by the use of personal passwords recognised by the computer or by digital signatures. Furthermore, the system should not accept any changes to data without concomitant entry of a reason or justification. In manual recording the entries made on a sheet of paper can be dated and signed to bear witness to the validity of data and to accept responsibility.

10.1.4 Therefore GLP wants to ensure that data safety and integrity remains the same in electronically as in manually recorded data, irrespective of how they were recorded, and that reconstruction of the way in which the final results and conclusions were obtained remains fully possible (Seiler, 2005). The Study Director must sign and date the final report to indicate acceptance of responsibility for the validity of all the data. (OECD, 1998).

11. Storage and retention of records and materials

Storage and retention of records and materials should be prepared appropriately. The following should be retained in the archives for the period specified by the appropriate authorities : the study plan, raw data, samples of test and reference items, specimens, and the final report of each study records of all inspections performed by the Quality Assurance Programme, as well as master schedules, records of qualifications, training, experience and job descriptions of personnel; records and reports of the maintenance and calibration of apparatus; validation documentation for computerised systems. In the absence of a necessitated retention period, the final arrangement of any study materials should be documented.The necessary documents for GLP regulations are given in Table 2.

GLP Regulations (Rules)	Documentation (Tools)
Organization and Personnel	Training records, CVs, GLP training
Facilities	Maintain adequate space/separation of chemicals from office areas
Equipment	Calibration, logbooks of use, repair, and maintenance; check freezers
Facility Operation	Standard operating procedures
Test, Control and Reference Substances	Chemical and sample inventory, track expiration dates, labeling
Records and Reports	Timely reporting, storage of raw data and reports

(Cobb., 2007).

Table 2. Documentation for GLP rules.

11.1 When samples of test and reference items and specimens are disposed of before the expiry of the necessitated retention period for any reason, this should be justified and documented. Material preserved in the archives should be indexed so as to facilitate storage and retrieval in a tidy way. Safe storage should be provided for all of the samples, test materials and the reports produced. Figure 4 shows the storage of test material

11.1.1 Only personnel authorised by management should have access to the archives. Movement of material in and out of the archives should be recorded appropriately.

11.1.2 Documentation should not be accepted only written documents but also the material generally related to the test facility. Quality Assurance is obliged to retain the respective records in a special archive. Therefore, management is responsible for providing archive facilities for the safe storage and recovery of study plans, raw data, final reports, samples of test items and specimens. Storage should be safe , therefore the design of, and environmental conditions in, these facilities should protect the archived material from ill-timed deterioration. Although it may be enough to archive paper raw data, study plans and final reports to support the necessary space under dry conditions, protected from fire, water and corrosive gases, more stringent conditions will be essential for the storage of tissue specimens from toxicology studies. Samples of the test and reference items should to be stored under the original conditions which were applied during the testing phase. Reconstruction of a study could only be possible if all documents, records and materials from this study can be made available in an unadulterated and unspoiled condition. Traceability in GLP means that there has to be a perfect nonstop line of evidence, chaining together the test item with the effects displayed by the test systems. GLP aims to minimise mistakes or mix-ups through extensive and specific labelling requirements. Documented information can be provided evidencing the application of the correct item in the stated amounts to the relevant test system.

11.1.3 The storage of records must enable their safekeeping for long periods of time without loss or deterioration. In order to encourage safe storage of data, restricted access is used to archive facilities and record the documents logged in and out to a limited number of staff. (Seiler, 2005).

11.1.4 During the conduct of multi-site studies, the temporary storage of materials should be carefully made. Such storage facilities should be safe enough and protect the integrity of their contents. When test site storage facilities are not adequate to satisfy GLP requirements, records and materials should be transferred to a GLP compliant archive. Test site management should ensure that adequate records are available to demonstrate test site involvement in the study. OECD, 1998).

Fig. 4. Storage of the test material in an organized order.

12. Summary

GLP regulations are summarized in Table 3.

GLP	Describes good practices for non-clinical lab studies that support research or marketing approvals for FDA-regulated products
GLP General Requirements	Appropriately qualified personnel Adequate resources Appropriate procedures for: -Sanitation, health precautions, clothing -Test protocol development, test methods -Data analysis, report development Appropriately qualified study director Quality assurance function
GLP Facilities Requirements	Suitable size, construction, segregation -Animal care -Animal supplies -Test & control products maintained in a secure area -Operating "suite" -Specimen & data storage
Equipment Requirements	Appropriately designed Adequate thru-put capacity Appropriately located Routinely maintained & calibrated
Standard Operating Procedures	Animal room prep Animal care Receipt, ID, storage, handling, mixing & sampling of test & control articles Test system observations Lab tests Handling of moribund or dead animals Necropsy or postmortem exams of animals Collection & ID of specimens Histopathology Data handling, storage & retrieval Equipment maintenance & calibration Transfer, proper placement & ID of animals

Reagents & Solutions	Adequate labeling -Identity -Concentration -Storage requirements -Expiration date
Test & Control Articles	Adequate characterization Proper receipt, storage, distribution When mixed with a carrier, adequate methods to confirm -Mixture uniformity -Article concentration -Article stability
Study Implementation	Written, approved protocol indicating test objectives & methods Study conducted in accordance with protocol Study monitoring to confirm protocol compliance Appropriate labeling of specimens by test system, study, nature & collection date Records of gross findings from postmortems available to pathologist for specimen histopathology
Standard data capture/recording requirements	-Legibility -Permanence -Accountability -Changes
Final report of results Study records & data methodically archived to facilitate expedient retrieval	-Study documents -Raw data -Specimens -Protocols -QA inspections -Personnel training & qualifications -Calibration & maintenance records
Records retention (shortest of):	-≥ 2 yr after FDA marketing clearance -≥ 5 yr after data submitted to FDA in support of marketing application -≥ 2 yr after Sponsor decision not to proceed with marketing application -Wet specimens hold as long as viable Records transferable with written FDA notification
Facility Disqualification	Grounds for disqualification: -Failure to comply with regulations & -Noncompliance adversely affects study validity & -Previous regulatory actions have been unsuccessful in modifying facility operations

(Clasby, 2005)

Table 3. GLP regulations.

12.1 "Good laboratory practice" can be considered as " essentially tidiness, cleanliness, hygiene and common sense." (CWIS, 2000).

12.1.1 Quality combination with the GLP rules will be the way that the laboratories will tend to select more in the next years. This will be the leading way to the evidence based laboratory results with a more trustworthy approach.

13. References

Clasby Ginger (2005). Good Laboratory Practice CFR 21 Part 58. A Review for OCRA US RAC Study Group September 2005. Available at :
http://www.google.com.tr/search?hl=tr&source=hp&q=A+Review+for+OCRA+US+RAC+Study+Group+September+2005+++&rlz=1W1ADFA_tr&aq=f&aqi=&aql=&oq

Cobb Jean (2007). GLP: Good Laboratory Practice for Field and Research. ALS 52 04 Available at:
http://www.docstoc.com/docs/18191459/Good-Laboratory-Practices

CWIS (2000), L1 – Good Laboratory Practice, Liverpool John Moores University, Campus Wide Information Service, Health and Safety Unit, Laboratory and Workshop Safety Notes Dalton, R (1999), Roche's Taq patent "obtained by deceit", rules US court, Nature (news) 402, 16 December, p. 709. Available at:
(www.cwis.livjm.ac.uk/hse)

Fox Arlene (2011). GLP Regulations vs. ISO 17025 Requirements: How do they differ? In Accreditation and Quality Assurance: Journal for Quality, Comparability, and Reliability in Chemical measurement. Volume 1/1996-volume 16/2011. DOI: 10.1007/s00769-011-0759-0. Available at:
https://springerlink3.metapress.com/content/mr20ux0343141g4k/resource-secured/?target=fulltext.pdf&sid=sbx4al45ojtfu3vvjzteu045&sh=www.springerlink.com

Gladney Lori, Osakwe Izabella, Ford Endia (2009). Good Laboratory Practices. Available at:
http://science.kennesaw.edu/~jhendrix/regs/GLP.ppt

OECD.(1998) OECD series on Principles of Good Laboratory Practice and Compliance Monitoring. Available at:
http://www.oecd.org/officialdocuments/displaydocumentpdf/?cote=env/mc/chem(98)17&doclanguage=en

Seiler Jürg P (2005) Good Laboratory Practice. The why and the how. ISBN 3-540-25348-3, Springer-Verlag Berlin Heidelberg, Printed in the European Union.

3

Evaluating Quality Control Decisions: A Simulation Approach

Mohamed K. Omar[1] and Sharmeeni Murugan[2]

[1]Nottingham University Business school Malaysia
[2]Faculty of Engineering & Technology Multimedia University
Malaysia

1. Introduction

Quality has become one of the core factors for almost all manufacturing and service companies that aim to achieve customer satisfaction. Therefore, improving quality is considered to be one of the efforts that companies consider a must to attain customer loyalty in today's complex global competitive environment. Studies concluded that any serious endeavour to improve quality will lead to an increase of cost of the product or service. Obviously, improving quality has its own costs. As a result, measuring cost of quality is very important as it provides valuable insights into the different cost of quality components. Thus, favourable returns on investment maybe achieved. For this fact, the quality cost concept was introduced and implemented in many manufacturing and service companies. The first model of cost of quality was introduced by Feigenbaum (1956) known as the P-A-F model which consists of prevention, appraisal and failure cost. Feigenbaum (1991) categorized the model into two major areas: costs of control (costs of conformance), and costs of failure of controls (costs of non-conformance), which since then used by numerous research studies (for example, Sumanth and Arora (1992), Burgess (1996), Purgslove and Dale (1995), Gupta and Campbell (1995), Chang et al. (1996), Sorqvist (1997)).

Most of the reported literature does not provide a single universal definition for cost of quality (Dale and Plunkett (1999)). However, cost of quality is usually best understood in terms of the sum of costs of conformance and the costs of non-conformance which was first introduced by Crosby (1979). Here, cost of conformance is known as the costs associate with quality requirement for achieving specific quality standards for a product or service. On the other hand, cost of non-conformance is known as the cost of failure to deliver the required standard of quality for a product or service. From the voluminous literature, one may categorized the cost of quality models into five generic models which are: P-A-F model, Crosby model, opportunity or intangible cost model, process cost models and ABC (activity based costing) model. Traditional cost accounting approaches typically used to measure cost of quality has been reported in the literature to have serious limitations when dealing with the components of intangible costs (see Son (1991), Chiadamrong (2003) and Sharma (2007)). As manufactures continue to improve their factories, they discover that existing cost measure systems should be updated and no matter how sophisticated and reliable these economic evaluation measure maybe, such problems still remain if unreliable cost information is not obtained as inputs for these economic evaluation, Chiadamrong (2003).

It is true to state that cost of quality modelling will provides more accurate approach to determine the cost involved in any quality control activities. However, the challenge does not end here. The cost of quality model must be used to determine the cost of improving activities associated with quality control strategies introduced to improve customer expectations. Therefore, a realistic cost of quality estimation could be determined that allows managers to show the economic benefit or otherwise of that specific quality control strategy. In other words, a quality control improvement strategy can only be justified if the increase in profitability is sufficient to cover the costs involved in the implementation. Once the cost of quality model is developed, a simulation can be used to determine the impact of any quality control strategy that a company wish to investigate. Among the strategies that manufacturing companies may consider investigating would be the allowable defect rate in some process or operations. In this case, simulation could be used to calculate the impact of defective rate of that operation on the overall profitability and productivity of the manufacturing system. Moreover, simulation could be used to study the system before and after some quality control improvement policy. Once the study is completed, a true picture about the cost of that policy could be determined as well as the impact of that policy on the overall defect rate.

This chapter is organized in the following manner; the literature review is presented in section 2 and followed by cost of quality model in section 3. Problem and solution methodology are presented in section 4. Experiment design and model verification, results and discussions are presented in sections 5 and 7 respectively. Finally, conclusions are presented in section 7.

2. Literature review

In its simplest definition of cost of quality, the American Society for Quality Control (ASQC (1971)) and the BS6143 Part 2 (1990) define cost of quality as the costs incurred in ensuring quality, together with the loss incurred when quality is not achieved. Feigenbaum (1956, 1961) introduces the so called PAF model in which cost of quality was classified into four components, prevention, appraisal and failure (internal and external) costs. In Plunkett and Dale (1987) survey, it is stated that literature suggests that most of the researchers use the PAF model for measuring cost of quality. However, Schiffauerova and Thomson (2006) reported that PAF model concept is not the only one since other models were found in the literature to be developed, discussed and used as detailed in Table 1. It is worth noting that Table 1 was originally developed by Schiffauerova and Thomson (2006) and updated by the authors.

The importance of cost of quality has been reported in many research works, Moyer and Gilmore (1979) reported that cost of quality could reach as high as 38% of sales and Albright and Roth (1992) estimated that the cost of quality may represents 30% of all manufacturing costs in the United States. Moreover, Harry and Schroeder (2000) asserted that most companies would find that cost of quality, if properly evaluated, falls somewhere between 15 and 25% of total sale-rather that 3-7% that often assumed. In other study, Giakatis et al. (2001) report that cost of quality represents considerable portion of company's total costs. More recently, Kent (2005) estimated that the turnover rates faced by companies are between 5 to 15% of the overall cost of quality. The benefits of implementing cost of quality system in any profitable organization has been reported extensively in the cost of quality literature, for example, Prickett and Rapley (2001) highlighted four common benefits that any organization is bound to gain from implementing cost of quality system :(1) it will be

Generic model	Cost/ activity categories	Examples of publications describing, analyzing or developing the model
P-A-F models	Prevention + appraisal + failure	Feigenbaum (1956), Purgslove and Dale (1995), Merino (1988), Chang et al.(1996), Sorqvist (1997), Plunkett and Dale (1988),Tatikonda and Tatikonda (1996), Bottorff (1997), Gupta and Campbell (1995), Burgess (1996), Dawes (1989), Sumanth and Arora (1992), Morse (1983), Weheba and Elshennawy (2004), etc.
Crosby's model	Conformance + non-conformance	Suminsky (1994) and Denton and Kowalski (1988)
Opportunity or intangible cost models	Prevention + appraisal + failure + opportunity	Sandoval-Chavez and Beruvides (1998) and Modarres and Ansari (1987)
	Conformance + non-conformance + opportunity	Carr (1992) and Malchi and McGurk (2001)
	Tangibles + intangibles	Juran et al.(1975)
	P-A-F (failure cost includes opportunity cost)	Heagy (1991) and Chiadamrong (2003)
Process cost models	Conformance + non-conformance	Ross (1977), Marsh (1989), Goulden and Rawlins (1995), Crossfield and Dale (1990)
ABC models	Value added + non-value added	Cooper (1988), Cooper and Kaplan (1988), Tsai (1998), Jorgenson and Enkerlin (1992), Dawes and Siff (1993) and Hester (1993)

Table 1. Generic cost models and cost categories adopted from Andrea and Thomson (2006).

able to focus upon areas of poor performance that need improvements, (2) it will have the opportunity to monitor the progress of ongoing improvement activities, (3) it will have an opportunity to plan for quality improvement and (4) it will be able to communicate better within the organization for improving the overall of the quality control.

Moreover, Schiffauerova and Thomson (2006) in their extensive literature review on cost of quality indicate that companies that use cost of quality programs have been quite successful in reducing cost of quality and improving the quality for the customer.

Although the impact of implementing cost of quality systems on increasing profit of any organization is obvious, Yang (2008) reported that the literature on cost of quality systems implementation indicates that most of companies do not know the true cost of their own quality. Despite that 82% of companies in the United Sates are involved in quality programs, only 33% actually compute the cost of quality (Harry and Schroeder, 2000), and in north-east England 66% of organization do not make use of quality costing as reported by Prickett and Rapley (2001). Some studies have highlighted reasons for the lack of implementing cost of quality systems in practice, Harry and Schroeder (2000) state that many significant quality related costs cannot be captured by most types of accounting systems. Chen and Yang (2002) related the difficulties to measure cost of quality to the fact that there is a lack of adequate methods for determining the financial consequences of poor quality. Moreover, Chiadamrong (2003) has concluded that there is a widespread belief that quality cost cannot be measured in practical terms because traditional cost accounting systems have not been adapted to quantify the value of quality. The need for quantifying cost of quality as stated by Yang (2008) and has been reported in the literature by many researchers, (Feigenbaum (1956); Juran (1952, 1989); Krishnan et al. (2000); Giakatis et al. (2001); Prickett and Rapley (2001), Chen and Yang (2002). However, there are evidences in the literature that clearly indicate that quantifying cost of quality has been neglected by most of organization (Harry and Schroeder (2000) and Omachonu et al. (2004)).

Quality performance is not something that can be readily altered in practice therefore one cannot test with the actual system. Most of the operations system are interconnected and subjected to both variability and complexity. Hence it is impossible to predict the performance of operation systems that are potentially subjected to variability, interconnectedness and complexity. Simulation has been used to overcome this problem and to investigate on the effect of quality loss financially and also examines the effect of different quality strategies on the financial aspect. Simulation provides a flexible technique to model an extensive array of issues that arise relating to quality and manufacturing. The flexibility of simulation methods permitted the invention of models with greater complexity than analytical techniques. Burgess (1996) constructed a simulation model based on system dynamics nature where the model was incorporated with the traditional P-A-F element. The model has facilitated the precise examination of the major relationships concerning conformance quality and costs at the organizational level. Gardner et al. (1995) examines the quality improvement in a manufacturing system by using simulation approach. The modelling was more complicated as it allows the defective parts to move along the assembly operations to examine the impact on the profitability and productivity.

Tannock (1995, 1997) emphasizes the significance of process capability in the selection of quality control strategy and revealed the economic advantages of control charting where special or assignable causes exist. Clark and Tannock (1999) investigate the use of simulation model to estimate the quality cost associated with multi manufacturing system setup and quality control strategies. This approach was validated and aligned with actual costs at a case study company. Although there are voluminous literatures written on cost of quality, very few literatures were written on tracing the invisible element of cost of quality and also the method in measuring the element of cost of quality.

The literature review presented so far indicates that estimating cost of quality is not a simple and straight forward issue. Moreover, the literature shows that there are several methods that can be used to determine cost of quality. Moreover, simulation stands as a favourite approach that might be used to investigate quality improvements justification in terms of profitability for a specific cost of implantation. Obviously, cost of quality model must first be developed and then simulation is followed to investigate the impact of some quality strategies and or improvements.

3. The cost of quality model

In this section of the chapter we, intend to provide a brief description of the model developed by Son and Lie (1991) and the way it was modified to be suitable for our research idea. First, the notations of the model are presented.

Notations

t	sampling interval
n	sample size per sampling
Θ	average number of assignable cause per hour
d_1	mean time spent to identify an assignable cause
d_2	mean time spent to correct an assignable cause
g	time required to sample a product
t_o	time inspection of the acceptance sampling
c_{as}	cost of investigating acceptance sampling
Γ	probability of process in control state 2
β	Probability of process in control state 2
τ	time taken for assignable cause to occurs
T1	in control period
T2	time period until the assignable is detected for the first time
T3	time required to investigate the true alarm
T4	time to correct the assignable cause

3.1 Description of the model

Son and Lie (1991) considered a small manufacturing system which consists of a machining area (sampling inspection) and a final inspection area (100% inspection) as indicated in Figure 1. However; they assume that 100% inspection is not always possible. They explained that a complete check of a component part may require the part removal from the fixture; the removal makes it difficult to realign the part to its original position.

Therefore, they assume that sampling inspection is assumed for component parts (in-process inspection), and 100% inspection for finished parts (final inspection). Throughout the machining process, the feature of the product quality is monitored by an x-bar chart which consists of a center line, lower, limit and an upper limit. A sample size in use is taken and inspected at a specific time interval. During machine process, when the sample mean fall outside of the control limit in the x-bar chart, an investigation is made during an average time period to check if the alarm is true or false. The false alarm occurs if the process is in

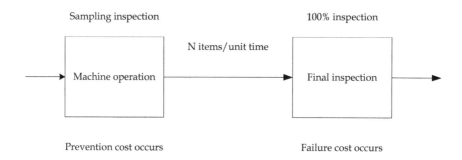

Fig. 1. Simplified manufacturing system.

control and on the other hand, the true alarm occurs if an assignable cause of specific magnitude makes the process out of control. In their modelling, first they calculate the cycle time, and then quality cost (prevention and failure) per cost was determined. In the next section, equations for all components are stated, obviously, readers interested in the detail steps of deriving these equations should referred to the original article.

3.1.1 Cycle time calculations
The assignable cause occurs according to the Poisson process with mean rate Θ. The period of the process is in control is represented by T_1. T_1 follows an exponential distribution with mean $1/\Theta$.
The expected length of a cycle;

$$E(T_c) = E(T_1) + E(T_2) + E(T_3) + E(T_4) \tag{1}$$

Which can be obtained using equation 2

$$= 1/\Theta + (t + t\Gamma/(1-\beta)-\tau) + (gn + d_1) + d_2 \tag{2}$$

3.1.2 Prevention cost per cycle
There are three types of prevention costs that occur at the machining area during a cycle of time. The first cost is associated with inspection works and denoted as (C_{p1}), the second cost is representing the cost of investigating the false alarms (C_{p2}), and the third cost is associated with adjusting the assignable cause (C_{p3}). Then, the expected prevention cost per cycle (C_p) can be determined by summing up these three components as in equation 3.

$$E(C_p) = E(C_{p1}) + E(C_{p2}) + E(C_{p3}) \tag{3}$$

3.1.3 Failure cost per cycle
The failure cost has three components; the first component is the cost of rework per cycle (C_{f1}), the second component is the scrap cost per cycle (C_{f2}) and the third component is the external failure cost during a cycle of time (C_{f3}). Then, the expected failure cost per cycle (C_F) can be determined by summing up these three components as in equation 4.

$$E(C_F) = E(C_{f1}) + E(C_{f2}) + E(C_{f3}) \tag{4}$$

3.1.4 Acceptance sample cost per cycle

Costs associated with sampling inspections prior entering the manufacturing system are not considered in the model developed by Son and Lie (1991). Our idea consists of the fact that companies often receive a shipment of material from a supplier and need to ascertain that the quality of the shipment. If it is impractical to inspect every item in the shipment, then a sampling plan is the (n, c) is used. In this plan, n items are chosen (without replacement) from a batch of shipped material. If c or fewer of the sample items are defective, then the batch is accepted; otherwise, the batch is rejected. Therefore, there will be a cost incurred during inspection of the sample before entering the manufacturing system. The expected number of samples entering the manufacturing system is n whereas t_o is representing the time taken to inspect the sample. Let (C_{aS}) be the cost of investigating acceptance sampling during a cycle of time. Then the expected cost associated with this activity is given by equation 5.

$$E\ (C_A) = C_{as}nt_o \tag{5}$$

Hence, the total cost of quality can be represented by summing up all the cost components (1) through (5) and is presented by equation 6.

$$E\ (C_Q) = E(C_P) + E(C_F) + E(C_A) \tag{6}$$

Where $E(C_P)$ is representing the prevention cost, $E(C_F)$ is representing the failure costs and $E(C_A)$ is representing the acceptance sampling costs.

4. The problem and solution methodology

The improvement made to the cost of quality model developed by Son and Lie (1991) that was described in section 3.1.4 adds acceptance sampling plans. Almost all quality control managers in the manufacturing firms develop and implement such plans. As a result an element of appraisal cost is incurred prior to the commencement of the production activities. Moreover, unlike the work reported by Son and Lie (1991) that consider a single stage manufacturing system, this research work considers a two-stage manufacturing system.

Therefore, the problem considered by this chapter can be described as: : a manufacturing system that consists of two stages, incoming raw materials inspection carried out according to some quality plan and as a result an element of appraisal cost in incurred prior to the commencement of the production activity. Once orders are realized by the company, raw materials are brought into the shop floor and at this stage; preventive and failure costs are incurred. In order to investigate the cost of quality for this manufacturing system, we have adopted the four strategies for inspection and removal of defectives across a range of detection rates reported by Gardner et.al (1995). The strategies are summarized in the following manner:

1. Inspection and removal of defectives based on acceptance sampling prior to assembly points
2. Inspection and removal of defectives at completion of finished product only,
3. Inspection and removal of defectives prior to assembly points,
4. Inspection and removal of defectives following every operations, and

The motivation for creating the above strategies (quality control decisions) and then using simulation tool is to investigate the impact of these strategies on cost of quality. One may conclude that the result of implementing any of these decisions is obvious. For example, the strategy that calls to carry out the inspection works at the final stage of the production and then removal of defectives will minimize the inspection costs, however, the increase of defectives at the final stage of manufacturing is inevitable. Moreover, inspection and removal of defectives prior to the assembly line will increase the quality prevention cost component and indirectly reduce the failure cost component. However, without the development of the cost of quality model and the simulation works, the magnitude of cost of quality could not be determined and the impact would not be known. It is not a secret that many firms sacrifices parts of the quality control steps under pressure to reduce cycle time.

Once the cost of quality model and its three components is developed, a simulation model using @Risk spreadsheet simulation software is developed for the two-stage process. The popularity using spreadsheet technology among practitioners justifies the use of the proposed @Risk simulation software.

5. Experiment design

Inspection time was considered as the main factor for calculating cost of quality. This is important as in real-life manufacturing systems; production line managers consider cycle time to be an important factor. Moreover, quite often, quality control activities or part of the activities would be sacrificed in order to attain a desired cycle time.

In order to simplify the experiment, it was decided to use the same fractions of defective rate to every operation in the manufacturing system for a given trial. The fractions of defective rate are divided into five categories: .001, 0.005, 0.010, 0.050 and 0.100. Every trial will run for about 10000 times of the simulation runs as indicated in Table 2.

Inspection and defective removal strategies	Fraction Detection per Operation				
	0.001	0.005	0.01	0.05	0.1
Acceptance sampling	10000 run	10000 run	10000 run	10000 run	10000 run
Completion of finish product	10000 run	10000 run	10000 run	10000 run	10000 run
Prior to assembly	10000 run	10000 run	10000 run	10000 run	10000 run
Following every operation	10000 run	10000 run	10000 run	10000 run	10000 run

Table 2. Experiment design for simulation run.

Therefore there will be twenty combinations of strategies and defective rates. All the combinations will run for 10000 iterations each. It is worth to note that during the design of the sampling acceptance procedure, it was assumed that the company would receives a batch of 10,000 items which require assembly activities using processes one and two respectively. Moreover, it is assumed that the quality engineers have developed a quality plan which is described as the (n, c) plan. In an (n, c) plan, n items are chosen (without replacement) from the batch of shipped material whereas c is the maximum number of defective items that a sample could have. If the number of the defective items in the sample is fewer or equal to c, then the batch is accepted otherwise the batch is rejected. For example, in process 1, a sample of 90 items will be inspected and the entire 1000 items are accepted if 0, 1, 2, 3, 4, or 5 defective items are found in the sample otherwise the batch is rejected.

On the other hand, for process 2, a sample of 40 items will be inspected and the entire batch will be accepted if 0, 1, 2 or 3 defective items are found in the sample otherwise the batch will be rejected. The idea here is to set the Actable Quality Level (AQL) to be equal to 1% and the Lot Tolerance Percentage Defective (LTPD) to be equal to 5%. Once AQL and the LTPD are set, a simulation run is carried out to determine the cost of quality associated with the quality control strategies (decisions) detailed in section 4 in this chapter. It is worth noting and as indicated earlier, the process under consideration consists of two processes. In order to avoid computational complexities; it was decided to apply and conduct simulation runs for each management strategies to each process separately and then combined the two processes to examine the overall findings. The numerical values which were used in the simulation model are portrayed in Tables 3 and 4.

Manufacturing system		Process 1	Process 2
	Symbol	Value	Value
Average number of assignable cause per hour	Θ	0.001	0.001
Lower specification level	LSL	0.79	0.99
Upper specification level	USL	0.81	1.01
Average height of the product population	μ	0.80	1.00
Standard deviation of the product population	σ	0.007	0.015
Magnitude of an assignable cause	δ	5	7
Value to decide control limit	q	3	7
Sampling interval	t	5	20
Sample size per sampling	n	90	40
Mean time spent to identify an assignable cause	d_1	0.5	0.5
Mean time spent to correct an assignable cause	d_2	2	2.5
Time required to sample a product	g	0.05	10

Table 3. Numeric values for processes 1 and 2.

Strategy 1. Inspection and removal of defectives based on acceptance sampling prior to assembly points.

This quality control strategy calls for the removal of defectives at the sampling stage and no quality control to be conducted after that stage. The total cost of quality associated with this strategy would be determined based on the following equation:

$$TCOQ_1 = C_{as}nt_o$$

Manufacturing system	Symbol	Process 1 Value	Process 2 Value
Number of products produced at the machining area per hour	N	40	20
Error rate of misclassifying a good part into bad	e_1	0.01	0.01
Error rate of misclassifying a bad part as good	e_2	0.005	0.005
Rate of restoring a defect to a good part	w	0.95	0.95
Fixed sampling cost per sampling interval	a_1	25	30
Variable sampling cost per unit product	a_2	8	5
Cost of investigating a false alarm	C_{fs}	80	80
Cost of correcting an assignable cause	C_{cr}	150	150
Cost of reworking a good part because of misclassification	C_g	100	100
Cost of reworking a defective part	C_b	200	200
Cost of scrapping a defective part that cannot be restored	C_s	75	75
Cost of dissatisfying a customer by selling a defective part	C_a	100	100

Table 4. Numeric values for processes 1 and 2.

Strategy 2. Inspection and removal of defectives at completion of finished product only.
This strategy calls for inspection of items among completion of the two processes. Therefore, it means that there will be no quality control during the process of making the products. Hence, the cost of quality that will be incurred in this strategy is meanly failure costs. However, it is assumed here that samples have to undergo visual inspection before the items enter into the manufacturing system. Therefore, the total costs of quality associates on this strategy are;

$$TCOQ_1 = E(C_A) + E(C_F) = E(C_A) + E(C_{f1}) + E(C_{f2}) + E(C_{f3})$$

Strategy 3. Inspection and removal of defectives prior to assembly points.
This strategy requires that items must be inspected and defectives are removed prior to assembly. In this case, since the inspection is done at the end of the process, the cost of quality is simply the prevention cost which is incurred during inspection of the material. However, parts with defects could be produced since there is no quality control during and at the end of the assembly. Therefore, there will be some failure costs which is required to convert defective items into good items. Moreover, inspection costs before the items enter the manufacturing system have to be considered. Therefore, the total costs of quality associates on this strategy are;

$$TCOQ_2 = E(C_A) + E(C_{p1}) + E(C_F) = E(C_A) + E(C_{p1}) + E(C_{f1}) + E(C_{f2}) + E(C_{f3})$$

Strategy 4. Inspection and removal of defectives following every operations.
This strategy calls for inspection and removal of defectives following every operation. Hence, the inspection will take place at the end of every operation to determine if there are defectives and perform the necessary corrections (rework). As a result, there will be an element of failure costs. Moreover, since the system includes inspection costs before items enter the manufacturing system, an additional inspection cost will be incurred. The total costs of quality associates on this strategy are;

$$TCOQ_3 = E(C_A) + E(C_{p1}) + E(C_{p2}) + E(C_F) = E(C_A) + E(C_{p1}) + E(C_{p2}) + E(C_{f1}) + E(C_{f2}) + E(C_{f3})$$

Each mathematical model that describe each strategy is developed in MS Excel spreadsheet for each process separately and then for the combinations of the two stages. The simulation software @Risk is imbedded and a simulation of 10000 iterations is carried out. This way, the impact of each of the strategy can be investigated and the impact on cost of quality can be determined.

6. Model verification, results and discussions

6.1 Model verification

A problem of model verifications arises when using simulation approach in decision evaluation. The literature referred to model verifications procedure developed by Naylor and Finger (1967) that ensures the developed model is free from any illogical error. In order to accomplish the task, the formulation and its parameters provided by Son and Lie (1991) were used to compute the prevention and failure costs reported by the authors. A simulation run was carried out and the results of the simulation model were compared with the results reported by Son and Lie (1991). The results are presented in Table 5.

Son and Lie (1991) results			The Simulation results		
Total length of a cycle, T_C (hours)	Prevention cost per cycle, C_P ($)	Failure cost per cycle, C_F ($)	Total length of a cycle, T_C (hours)	Prevention cost per cycle, C_P ($)	Failure cost per cycle, C_F ($)
$E[T_1]=$ 100.00	$E(C_{P1})=$ $312.83	$E(C_{F1})=$ $3292.96	$E[T_1]=$ 100	$E(C_{P1})=$ $312.83	$E(C_{F1})=$ $3292.96
$E[T_2]=$ 4.28	$E(C_{P2}) =$ $2.63	$E(C_{F2})=$ $214.21	$E[T_2]= 4.28$	$E(C_{P2}) =$ $2.63	$E(C_{F2})=$ $214.21
$E[T_3]=$ 0.55	$E(C_{P3})=$ $100.00	$E(C_{F3})=$ $32.29	$E[T_3]=$ 0.55	$E(C_{P3})=$ $100.00	$E(C_{F3})=$ $32.29
$E[T_4] =$ 2			$E[T_4] =$ 2		
$E[T_C]=$ 106.83	$E[C_P]=$ $415.46	$E[C_F]=$ $3539.46	$E[T_C]=$ 106.83	$E[C_P]=$ $415.46	$E[C_F]=$ $3359.46

Table 5. Comparison results.

Table 5 indicates that the results obtained from the simulation model match exactly the results reported by Son and Lie (1991). The results from this finding indicate that the simulation model is an accurate presentation of that reported by Son and Lie (1991). Hence, it is concluded that the model is ready for investigation the evaluating the quality control decisions.

6.2 Results and discussions
6.2.1 Acceptance sampling prior to assembly points
As indicted earlier, the quality control engineers develop quality control plan.
In this inspection as mentioned in section 4, the quality control engineers have developed a quality control plan based on the concept of producer's risk which is related to acceptable quality level (AQL) and the consumer's risk which is related to tolerance percentage defective (LTPD).
The summary of the statistics that resulted from the simulations runs shows that the mean was found to be equal to 0.994 with a tile of 15% and the maximum is equal to 1 with a tile of 5%. Since it was concluded that the sample is good, since 99.94% of the sample is within 1% of defectives, then the items are considered to be acceptable. On the other hand, the remaining 0.06% of the sample contributes to produce risk. This means that the plan has a risk of losing 0.06% of good sample. It is worth noting that the skewness has a negative value that is -40.7941969 indicating that distribution of the sample exceedingly to the left.

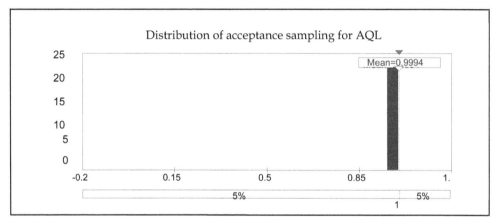

Fig. 2. Distribution of acceptance sampling for AQL.

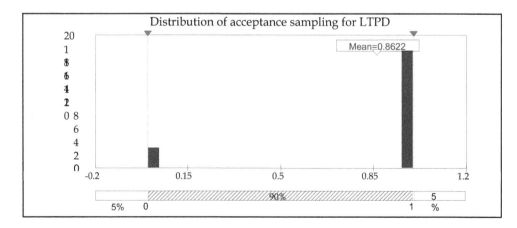

Fig. 3. Distribution of acceptance sampling for LTPD.

Figure 2 was constructed to display the summary statistics, the mean value for the acceptable quality level (AQL) over the tile percentage. From the chart it could be seen that the mean of the sample is equal to 0.9994 which very close to the value of 1 which indicate that the risk is equal to 0.00006.

As for the risk associated with low tolerance percentage defect (LTPD), the statistics summary results indicate that the mean has a value of 0.8622 which says that 86.22% of the sample size has defects and would be accepted as good parts. Therefore, such situation indicates that the risk would be about 86.22% of the sampling plan. Figure 3 summarizes the summary statistics. In Figure 3, the distribution is skewed towards the right represents the lot tolerance percentage defects with a mean of 86.22%.

6.2.2 Inspection upon completion of assembly

In this case inspection and removal will took place upon completion of assembly activities. Simulation runs resulted in obtaining the cost of quality associated with strategy of inspection upon completion of assembly activities at each level of defects rate per operation is presented in Table 6.

Cost of Quality (x 1000)				
Defective rate per operation (%)	Time interval (t, hr) and time inspection of a product (g, hr), Cost is in MU=Monetary Unit			
	t=20 g=10	t=17 g=9	t=15 g=8	t=10 g=7	t=8 g=6
0.001	3558.95	3398.49	3239.98	3075.65	2917.15
0.005	1989.88	1829.28	1670.68	1506.19	1347.65
0.01	1794.33	1633.55	1474.85	1310.16	1151.55
0.05	1642.42	1480.26	1320.76	1154.48	995.41
0.10	1628.56	1464.93	1304.54	1136.43	976.81

Table 6. Cost of quality at different values of defect rate, time interval and inspection time.

Table 6 reveals the total cost of quality is decreasing as we move towards the right side of the table. The reason for that is the fact that the more time is allocated for inspection work, the more of the defects will be found. This finding could be seen for example when considering the total cost at 0.001 defects rate at interval time of 20 hours and inspection time for 10 hours. Considering that example reveals that the cost has decrease by 27%. Moreover, Table 6 indicates that the maximum impact on total cost of quality resulted from implanting this strategy occurred at defect rate 0.001 with time equal to 20 hours and inspection time of 10 hours. On the other hand, the minimum cost of quality occurs at a defect rate of 0.10 with 8 hours interval time of inspection. The total cost of quality behaviour at various interval times for each defect rate is presented in Figure 4.

Figure 4 reveals that there is a large gap between the curve for the defects rate at 0.001 and the curve represents the defect rate at 0.005. As it could be seen, there is a drastic drop and the drop is due to the increase of the defects that will cause the time taken to inspect the defect is shorter and thus the time to correct the assignable cause shorter. Hence, it will reduce the total cost of quality in this situation compared to the lower defect rate. Moreover, the difference of the total cost of quality at defect rate of 0.05 and defect rate of 0.10 is small. Therefore, the curves represent this defect rate is overlapping with each other's and the curve that represents the defect rate of 0.10 has the smallest total cost of quality.

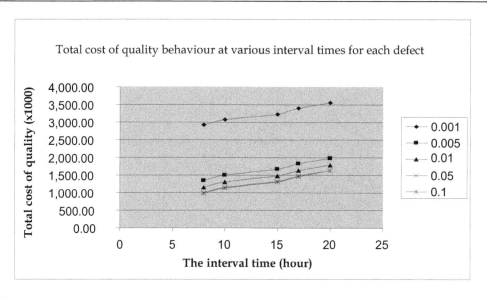

Fig. 4. Cost of quality behaviour at various interval times for each defect rate.

6.2.3 Inspection prior to assembly operations
This quality control strategy calls for inspection to be carried out at the commencement of the assembly process; the cost of quality at various defective rates and inspection interval times is presented in Table 7.

Cost of Quality (x 1000)					
Defective rate per operation (%)	Time interval (t, hr) and time inspection of a product (g, hr), Cost is in MU=Monetary Unit				
	t=20 g=10	t=17 g=9	t=15 g=8	t=10 g=7	t=8 g=6
0.001	3073.71	2937.75	2803.23	2669.58	2539.01
0.005	1715.11	1577.40	1441.37	1301.44	1166.24
0.01	1545.79	1407.72	1271.42	1130.55	994.72
0.05	1414.24	1274.82	1137.69	994.84	858.15
0.10	1402.23	1261.52	1123.60	979.09	841.87

Table 7. Cost of quality at different values of defect rate, time interval and inspection time.

Table 7 reveals that the maximum total cost of quality occurs at defect rate of 0.001 and time interval of 20 hours and inspection time of 10 hours. On the other hand, the minimum total cost of quality occurs at defective rate of .10, time interval of 8 hours and inspection time of 6 hours. As it could be noticed that as the time of inspection become shorter, total cost of quality is reduced. The relationship between total cost of quality, the time interval and defective rate is presented in Figure 5.

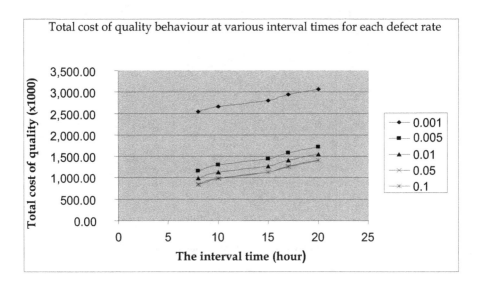

Fig. 5. Cost of quality behaviour at various interval times for each defect rate.

6.2.4 Inspection following every assembly operations

This strategy calls for inspection works to be carried out at the end of the process for every operation. The cost of quality at various defective rates and inspection interval times is presented in Table 8.

Cost of Quality (x 1000)					
Defective rate per operation (%)	Time interval (t, hr) and time inspection of a product (g, hr), Cost is in MU=Monetary Unit				
	t=20 g=10	t=17 g=9	t=15 g=8	t=10 g=7	t=8 g=6
0.001	3062.25	2924.25	2787.93	2646.61	2510.30
0.005	1712.85	1574.73	1438.34	1296.88	1160.53
0.01	1544.67	1406.40	1269.92	1128.29	991.88
0.05	1414.03	1274.57	1137.40	994.40	857.61
0.10	1402.11	1261.39	1123.45	978.88	841.60

Table 8. Cost of quality at different values of defect rate, time interval and inspection time.

Table 8 reveals that the maximum total cost of quality occurs at defective rate of 0.001, interval time of 20 hours and inspection time of 10 hours. On the other hand, the minim total cost of quality occurs at defective rate of 0.10, time interval of 8 hours and inspection time of 6 hours.. The relationship between total cost of quality, the time interval and defective rate is presented in Figure 6.

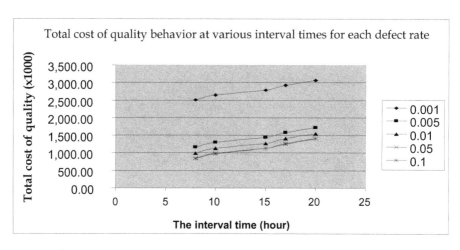

Fig. 6. Cost of quality behaviour at various interval times for each defect rate.

Figure 6 reveals that as defective rate per operation is increased, the total cost of quality decreases. This is because as the defective rate is increased, the time taken to detect the defect becomes shorter. Hence, the time taken to convert the process out of control to in control state will be shorter and finally the impact of cost of quality will be smaller.

In order to provide a good picture to the reader, Figure 7 was developed and a graphical presentation that shows the relationship between total cost of quality, and detection rate per operation.

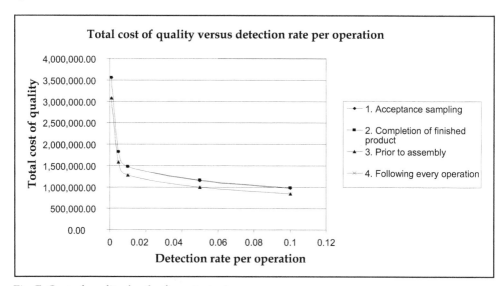

Fig. 7. Cost of quality for the four strategies.

Figure 7 indicates that total cost of quality is at its maximum value when the detection rate is zero, and then steadily decreased as the window of inspection increases. Moreover, inspection following every assembly operations has the lowest total cost of quality. This is because as the appraisal cost increases, the failure costs will simultaneously reduced since less defect rate occurs and hence less items is sent for rework.

6.2.5 Investigation of strategies combination

Since simulation modelling can provide a window of opportunity to investigate important quality control decisions and their impact on cost of quality. It is thought that the quality control management may wish to consider some combination of strategies for each process rather than having a single strategy for the whole assembly process. In this case, one has to determine the possible quality strategies combinations that could be considered for the simulation run. Let R be the possible strategies combination outcome, then we would have:

$$R = \sum S_{i,j} \qquad (7)$$

Where

i = A strategy in process 1 i = 1, 2, 3, 4 and j = A strategy in process 2 j = 1, 2, 3, 4

Inspection and defective removal strategies	Fraction Detection Rate per Operation				
	0.001	0.005	0.01	0.05	0.1
$S_{1,1}$	4,845.07	2,137.34	1,651.57	1,224.13	1,023.91
$S_{1,2}$	4,854.02	2,141.85	1,655.16	1,226.89	1,026.21
$S_{1,3}$	4,368.78	1,889.97	1,451.73	1,067.25	891.27
$S_{1,4}$	4,357.32	1,887.30	1,450.23	1,066.82	891.01
$S_{2,1}$	4,848.23	2,138.03	1,651.92	1,224.21	1,023.93
$S_{2,2}$	4,857.18	2,142.53	1,655.51	1,226.97	1,026.23
$S_{2,3}$	4,371.95	1,890.65	1,452.08	1,067.32	891.30
$S_{2,4}$	4,360.48	1,887.98	1,450.58	1,066.89	891.03
$S_{3,1}$	4,622.03	2,090.06	1,630.57	1,217.32	1,023.83
$S_{3,2}$	4,630.98	2,094.56	1,634.16	1,220.08	1,026.13
$S_{3,3}$	4,145.74	1,842.68	1,430.73	1,060.44	891.20
$S_{3,4}$	4,134.28	1,840.00	1,429.23	1,060.01	890.93
$S_{4,1}$	4,472.97	2,052.62	1,605.53	1,209.65	1,016.16
$S_{4,2}$	4,481.93	2,057.12	1,609.12	1,212.41	1,018,.47
$S_{4,3}$	3,996.69	1,805.25	1,405.69	1,052.76	883.53
$S_{4,4}$	3,985.23	1,802.57	1,404.19	1,052.33	883.26

Table 9. Total cost of quality for the assembly system.

Table 9 shows all possible quality strategies combinations with their respective cost at each selected fraction detection rate for each operation. As it could be seen from Table 9, the combination of {2, 2} strategies have caused the system to incur the largest cost of quality. This is the case when the management decided to use the quality strategy in which quality control inspection work is only to be carried at the end of the production runs. On the other hand, one can see from Table 9 that that adopting combination of {4, 4} strategies have caused the system to incur the least cost of quality. As it could be easily seen that {4, 4} strategies calls for quality control works to be carried after each operation and as a result, completely eliminating the failure costs. There are many important management implications illustrated in all the examples presented in this chapter. First, quality costs are very large when quality activities time window is sacrificed to reduce the total cycle time. Secondly, failure costs are very large and managers should completely avoid these costs since there is no trade off that exists with these costs. Another important issue is the fact that using simulation to measure and understand cost of quality has provide managers with opportunity to rank their process in terms of cost of quality and the cost consequences that resulted from adopting a specific set of strategies .

7. Conclusions

In this chapter, an analytical model reported in the literature for cost of quality computations was considered and modified to include an important component of cost of quality. The model is then used to develop a simulation model for a two-stage manufacturing system. Moreover, quality control strategies common in the manufacturing community were used in the simulation works to investigate their impact on cost of quality. The results indicate that some of these practiced strategies when combined with the detection periods will significantly increase cost of quality. Furthermore, the results indicate simulation works can be used to understand and measure cost of quality. An interesting venue for further research is to use simulation to investigate a real-industrial application to investigate the cost of quality; the authors are currently considering such idea.

8. References

Albright, T.L. and Roth, H. P (1992) .The measurement of quality cost: an alternative paradigm, *Accounting Horizons,* June, 15, 1992.

ASQC . (1971) Quality costs, what and how? Milwaukee: WI: ASQC Quality Press.

Bottorff, D. L.(1997). CoQ systems: the right stuff, *Quality Progress*, March, 33.

BS 6143 Part 2. (1990) Guide to determination and use of quality-related costs. London: BSI.

Burgess, T.F. (1996). Modeling quality cost dynamics, *International Journal of Quality & Reliability Management*, 13(3), .8-26.

Carr, L.P. (1992). Applying cost of quality to a service business, *Sloan Management Reviews,* 33(4), 72-78.

Chang, S.J., Hyun, P.Y. and Park, E. H. (1996). Quality costs in multi-stage manufacturing systems, *Computers & Industrial Engineering,* 31(1-2), 115-118.

Chen, C.C. and Yang, C.C. (2002). Cost-effectiveness based performance evaluation for suppliers and operations, *Quality Management Journal*, 9(4), 59–73.

Chiadamrong, N. (2003). The development of an economic quality cost model, *Total Quality Management and Business Excellence*, 14(9), 999-1014.

Clark, H.J and Tannock, J.D.T. (1999), The development and implementation of a simulation tool for the assessment of quality economics within a cell-based manufacturing company, *International Journal of Production Research*, 37, 979-995.

Cooper, R. (1988). The rise of activity-based costing – Part I: What is an activity-based cost system?, *Journal of Cost Management*, 2(2),.45-54.

Cooper, R. and Kaplan, R. S. (1988). Measure costs right: Make the right decisions, *Harvard Business Review*, 66(5), .96-103.

Crosby, P. B. (1979). Quality is free, New York: McGraw-Hill.

Crossfield , R. T. and Dale, B. G. (1990). Mapping quality assurance systems: a methodology, *Quality and Reliability Engineering International*, 6(3), 167-178.

Dale, B.G. and Plunkett, J. J. (1999). Quality Costing, 3rd ed., Gower Press, Aldershot.

Dawes, E.W. (1989). Quality costs-new concepts and methods, quality costs: ideas & applications, in Campanella, J. (Ed.), *Quality Costs: Ideas and Applications*, Vol. 2(pp.440.), ASQC Quality Press, Milwaukee, WI.

Dawes, E.W. and Siff, W. (1993). Using quality costs for continuous improvement, *ASQC Annual Quality Congress Transactions*, 444-449.

Denton, D.K. and Kowalski, T.P. (1988). Measuring nonconforming costs reduced manufacturer's cost of quality in product by \$200 000, *Industrial Engineering*, 20, 36-43.

Feigenbaum, A. V. (1956). Total quality control, *Harvard Business Review*, 34(6), 93-101.

Feigenbaum, A. V. (1961). Total Quality Control, McGraw-Hill Inc., New York, USA.

Gardner, L.L, Grant, M.E. and Rolston, L.J. (1995). Using simulation to assess costs of quality, *Proceedings of the Winter Simulation Conference*.

Giakatis, G., Enkawa, T. & Washitani, K. (2001). Hidden quality costs and the distinction between quality cost and quality loss, *Total Quality Management*, 12(2), 179-190.

Goulden, C. and Rawlins, L. (1995). A hybrid model for process quality costing, *International Journal of Quality & Reliability Management*, 12(8), 32-47.

Gupta, M. and Campbell, V. S. (1995). The cost of quality, *Productions and Inventory Management Journal*, 36(3), 43-49.

Harry, M.J. and Schroeder, R. (2000). Six Sigma: The breakthrough management strategy revolutionizing the world's top corporations, New York: Doubleday, Random House.

Heagy, C.D. (1991). Determining optimal quality costs by considering costs of loss sales, *Journal of Cost Management for the Manufacturing Industry*, Fall, 67-71.

Hester, W.F. (1993). True quality cost with activity based costing, *ASQC Annual Quality Congress Transactions*, 446-453.

Jorgenson, D.M. and Enkerlin, M.E. (1992). Managing quality costs with the help of activity based costing, *Journal of Electronics Manufacturing*, 2, 153-160.

Juran, J .M. (1952). Quality Control Handbook (1st ed.). New York: McGraw-Hill.

Juran, J. M. (1989). Juran on Leadership for Quality. New York: Free Press.

Juran, J.M., Gryna, F.M. and Bingham, R. (1975). Quality Control Handbook, 3rd edition, New York, McGraw-Hill.

Kent, R. (2005). Manufacturing strategy for window fabricators 14 – the cost of quality, Tanagram Technology, available at: www.tanagram.co.uk.

Krishnan, S.K., Agus, A. and Husain, N. (2000). Cost of quality: The hidden costs, *Total Quality Management*, 11(4, 5 & 6), 844–848.

Malchi, G. and McGurk, H. (2001). Increasing value through the measurement of the cost of quality (CoQ) – A practical approach, *Pharmaceutical Engineering*, 21(3), 92-95.

Marsh, J., (1989).Process modeling for quality improvement, *Proceedings of the Second International Conference on Total Quality Management*, IFS publication, Bedford, 111-121.

Merino, D. N. (1988). Economics of quality: Choosing among prevention alternatives, *International Journal of Quality & Reliability Management*, 5(7), 13-23.

Modarres, B. and Ansari, A. (1987). Two new dimensions in the cost of quality, *International Journal of Quality & Reliability Management*, 4(4), 9-20.

Morse, W. J. (1983). Consumer product quality control cost revisited, *Measuring Quality Costs, Cost and Management,* July/August, 16-20.

Moyer, D.R. and Gilmore, H.L. (1979) Product conformance in the steel foundry jobbing shop, *Quality Progress,* 12 (5),. 17-19.

Naylor, T.H., and Finger, J.M. (1967), Verification of computer models, *Management Science,* 14, pp92.

Omachonu, V.K., Suthummanon, S. And Einspruch, N. G. (2004). The Relationship Between Quality and Cost of Quality for a Manufacturing Company, *International of Quality & Reliability Management,* 21 (3), 277-290.

Plunkett, J.J. and Dale, B. G. (1988).Quality costs: A critique of some 'economic cost of quality' models, *International Journal of Production Research,* 26(11), 1713-1726.

Plunkett, J.J., Dale, B. G. (1987). A review of the literature on quality-related costs, *International Journal of Quality & Reliability Management,* 4 (1), .40-52.

Prickett, T.W. and Rapley, C. W. (2001). Quality costing: A study of manufacturing organizations Part 2: Main survey, *Total Quality Management,* 12(2), 211–222.

Purgslove, A.B. and Dale, B. G. (1995). Developing a quality costing system: Key features and outcomes, Omega: *International Journal of Management Science,* 23 (5), 567-575.

Ross, D.T. (1977). Structured analysis (SA): A language for communicating ideas, *IEEE Transactions on Software Engineering,* 3(1), 16-34.

Sandoval-Chavez, D.A. and Beruvides, M. G. (1998). Using opportunity costs to determine the cost of quality: A case study in a continuous-process industry, *Engineering Economist,* 43(2), 107-124.

Schiffauerova, A. and Thomson, V. (2006). A review of research on cost of quality models and best practices, *International Journal of Quality and Reliability Management,* 23 (6), 647-669.

Sharma, J. K. (2007). Business Statistics, 2nd edition, Pearson Education India.

Son, Y.K and Lie, F.H. (1991), A method of measuring quality costs, *International Journal of Production Research,* Vol. 29, pp. 1785-1794.

Sorqvist, L. (1997). Effective methods for measuring the cost of poor quality, *European Quality,* 4 (3), 40-42.

Sumanth, D. J., and Arora, D. P. S.(1992). State of the art on linkage between quality, quality costs and productivity, *International Journal of Materials and Product Technology,* 7(2), 150-169.

Suminsky Jr., L. T. (1994). Measuring cost of quality, *Quality Digest,* 14 (3), 26-32.

Tannock, J.D.T. (1995). Choice of inspection strategy using quality simulation, *International Journal of Quality & Reliability Management,* Vol. 12, No. 6, pp. 75-84

Tannock, J.D.T. (1997). An economic comparison of inspection and control charting using simulation, *International Journal of Quality & Reliability Management,* Vol. 14, No. 7, pp. 687-700.

Tatikonda, L.U. and Tatikonda, R.J. (1996). Measuring and reporting the cost of quality, *Production and Inventory Management Journal,* 37(2), 1-7.

Tsai, W. H. (1998). Quality cost measurement under activity-based costing, *International Journal of Quality and Reliability Management,* 15(7), 719-752.

Weheba, G. S. and. Elshennawy, A. K. (2004). A revised model for the cost of quality, *International Journal of Quality & Reliability Management,* 21(3), 291-308.

Yang, C.C. (2008). Improving the definition and quantification of quality costs, *Total Quality Management,* 19(3), March, 175 – 191.

Part 2

Evaluating Analytical Data

4

Patient Satisfaction with Primary Health Care Services in a Selected District Municipality of the Eastern Cape of South Africa

N. Phaswana-Mafuya[1,2], A. S. Davids[1], I. Senekal[3] and S. Munyaka[3]
[1]Human Sciences Research Council, Port Elizabeth,
[2]Office of the Deputy Vice Chancellor: Research and Engagement, Nelson Mandela
Metropolitan University, Port Elizabeth,
[3]University of Fort Hare,
South Africa

1. Introduction

Traditionally, decisions about health services were made on the basis of health-provider and health authorities' views on what is in the best interest of the patient. This was based on a view that members of the general public lack the technical knowledge to make fully informed decisions themselves. Currently, the use of patient satisfaction surveys (PSS) in developing countries is advancing. Professionals have recognized that a systematic and consumer oriented perspective toward patient viewpoints about the level of care can result in feedback useful for promoting higher quality standards of patient care (Dağdeviren & Akturk 2004; Newman et al. 1998; Peltzer 2009).

Patient satisfaction surveys are seen as a means of determining patients' views on primary health care (PHC) (Ajayi, Olumide & Oyediran 2005; Andaleeb 2001; Campbell, Ramsay & Green 2001). These surveys are increasingly being promoted as a means of understanding health care service quality and the demand for these services in developing countries (Glick 2009) for various reasons. First, they highlight those aspects of care that need improvement in a health care setting (Ajayi, Olumide & Oyediran 2005; Muhondwa *et al.* 2008; Newman *et al.* 1998). Second, they are simple, quick and inexpensive to administer. Third, they are critical for developing measures to increase the utilization of PHC services. Fourth, they can help to educate medical staff about their achievements as well as their failures, assisting them to be more responsive to their patients' needs. Fifth, they allow managerial judgment to be exercised from a position of knowledge rather than guesswork in the important task of managing public expectations and resources (Glick 2009).

The South African government also endorses the centrality of consumers in service delivery. The White Paper on Transforming Public Services of 1997 (Department of Public Service and Administration 1997) and the Department of Health's policy on quality in health care (Department of Health 2007) state that public services need to respond to customers' needs, wants and expectations. Feedback from consumers is required in terms of experiences of health services – quality of care received. Feedback from customers will not only improve knowledge of decision makers, but will also facilitate more improved prioritization, improved strategic resource allocation and improved value for money. It will also serve as a platform for providing better services to citizens.

Against this background, a patient satisfaction survey with PHC services was conducted in a selected district of the Eastern Cape.

2. Methods

2.1 Design and setting

A cross-sectional descriptive design was employed to collect data among patients visiting 12 clinics in a selected health district of the Eastern Cape of South Africa in 2009. The majority of the South Africans are dependent on the public health sector, with only 15% of the citizenry belonging to a private medical aid scheme (McIntyre, 2010). In the Eastern Cape, private medical aid covers only 10.9% of the province's population and less than 7% of South Africa's private and independent hospitals, are located in the Eastern Cape (Hospital Association of South Africa, 2011). The current study focused on public health services (the main provider of health care) in a selected district in the Eastern Cape. We did not ask about private health care utilization.

The public health system of the Eastern Cape consists of 817 clinics, 81 hospitals and 18 community health care centres. The core norms, set by the National Department of Health in South Africa, for primary health care services are indicated in Table 1. Statistics South Africa estimated that the mid-year population for Eastern Cape in 2010 was 6743800, about 13,5% of the estimated total population of South Africa. Persons under the age of 15 years constitute 32.8% of the total population and the economically active population (15-64 years) is 61.2% of total population of the Eastern Cape. For the period 2006 to 2011 it is estimated that approximately 211 600 people will migrate from the Eastern Cape to other provinces of the country (Statistics South Africa, 2010). The burden of disease study for the year 2000, estimated that South Africans suffer from poverty-related diseases and conditions, emerging chronic diseases, injuries and HIV/AIDS, and differences in morbidity and mortality between socioeconomic groups (Bradshaw et al., 2003). The Eastern Cape had an estimated 23.6% unemployment rate at the end of the second quarter of 2009. This drives levels of poverty in the province, as those deemed to be living in poverty was 3564504, nearly 53% of the 2010 midyear population. The corresponding figure for South Africa is 38%, making the province one of the poorest in the country (ECSECC, 2011).

2.2 Sample and procedure

A purposive sample of 836 out of 939 patients (89% response rate), visiting 12 primary care facilities in a selected district of the Eastern Cape of South Africa were interviewed while exiting the clinic. Patients aged 18 years or above were considered eligible, provided that they were able to understand and respond to the interview questions. Patients were interviewed face to face by trained interviewers in their preferred language in five consecutive days per clinic. Two fieldworkers and one fieldwork coordinator (with at least a high school certificate) were trained per clinic. Two of the four local fieldworkers conducted the interviews in the respective clinics, while the other two served as a reserve. A clinic nurse supervised and coordinated the fieldwork process in the clinic where s/he was based in. Ethics approval for the study protocol was obtained from the University of Fort Hare's Research Ethics Committee and permission to conduct the study was received from the Eastern Cape Department of Health.

Core Norms	Core Services
• The clinic renders comprehensive integrated PHC services using a one-stop approach for at least 8 hours a day, five days a week. • Access, as measured by the proportion of people living within 5km of a clinic, is improved. • The clinic receives a supportive monitoring visit at least once a month to support personnel, monitor the quality of service and identify needs and priorities. • The clinic has at least one member of staff who has completed a recognised PHC course. • Doctors and other specialised professionals are accessible for consultation, support and referral and provide periodic visits. • Clinic managers receive training in facilitation skills and primary health care management. • There is an annual evaluation of the provision of the PHC services to reduce the gap between needs and service provision using a situation analysis of the community's health needs and the regular health information data collected at the clinic. • There is annual plan based on this evaluation. • The clinic has a mechanism for monitoring services and quality assurance and at least one annual service audit. • Community perception of services is tested at least twice a year through patient interviews or anonymous patient questionnaires.	• Women's Reproductive Health • Integrated Management of Childhood Illness • Diseases prevented by Immunisation • Adolescent and Youth Health • Management of Communicable Disease • Control of Cholera, diarrhoeal disease and dysentery • Helminths • Sexual Transmitted Diseases (STD)and HIV/AIDS • Malaria • Rabies • Tuberculosis • Leprosy • Prevention of Hearing Impairment due to Otitis Media • Rheumatic Fever and Haemolytic Streptococcal Infection • Trauma and Emergency • Oral and Mental Health • Treatment and support of victims of Sexual Offenses, Domestic Violence and Gender Violence • Substance Abuse • Chronic Diseases, Diabetes, Hypertension • Geriatric care • Rehabilitation Services

Source: Department of Health, South Africa

Table 1. The core norms and services for primary health care PHC set by the NDOH.

2.3 Data collection method

A patient satisfaction questionnaire adapted from the one developed by the Health Systems Trust in 2004 was used. Only slight changes were made to the questionnaire in collaboration with the Eastern Cape Department of Health to allow for cross-comparisons with earlier patient surveys that have been undertaken within the Eastern Cape Province using the same questionnaire. Further, some questions were asked on demographics, health status, reason for health visit, and health care utilization. The questionnaire was translated from English into Afrikaans and Xhosa. The Xhosa and Afrikaans versions were developed using back-translation methods (Brislin 1970). The procedure entailed having two native-speakers of the target languages independently do a back-translation. Discrepancies were arbitrated by a third consultant, and solutions were reached by consensus. The translated questionnaire underwent pilot-testing.

2.4 Measures
The questionnaire included demographics and eight domains, each having several items on a 5-point likert scale: Strongly Agree=5; Agree=4; Unsure=3; Disagree=2; and Strongly Disagree=1

2.5 Data analysis
Data was captured on SPSS version 17.0 and analysed. Frequency distributions of domain items were made and positive responses (Agree and Strongly agree) were grouped and are presented. Cross tabulations of domain items by gender were made. Chi square tests were performed to determine the relationship between each domain item and gender.

2.6 Limitations
Response biases introduced through the methodology of using exit interviews might act as filters and influence patient satisfaction ratings. For example, exit interviews automatically select out those who do not have access to public health facilities, but would otherwise have used services. In addition, using exit interviews in health facilities identified by the sub-district officials, means that respondents were purposively selected. Non-randomisation in the selection of respondents means that results are more difficult to generalise to a feeder population around a health facility. The study compensated for this limitation by collecting data from each facility over a week during a period of normal use and through achieving a high number of respondents. A further limitation is that the existing PSS methodology does not enable the relationship between aggregate satisfaction scores and changes in health status of populations to be explored.

3. Results

3.1 Sample characteristics
The majority of the respondents were African (50.9%), female (72.9) and unemployed (56.4%) with a mean age of 39.4 years. Only 5.5% of the respondents indicated that they had enough money to meet their basic needs for most of the time. Almost 85% had some form of formal education.

3.2 Utilization of health services
Most respondents visited clinics more frequently (i.e. about 7 times within 12 months) compared to private doctors (i.e. about 1.58 times), hospitals (i.e. about 1.14 times) and traditional healers were list visited (0.23 times). The main reason for visiting the health facility was to get treatment (41.8%), followed by suffering from non-communicable diseases (NCDs) (11.5%) and family planning (10.1%).

Mean Age: M (SD)	39 years (14.91)
Gender	**N (%)**
Male	230 (29.9)
Female	674 (72.9)
Race	**N (%)**
African	469 (50.9)
White	26 (2.8)
Indian	10 (1.1)
Coloured	375 (40.7)
Other	39 (4.2)
Occupation	**N (%)**
Employed	335 (37.1)
Not Employed	509 (56.4)
Other	59 (6.4)
Highest Level of Education	**N (%)**
None	139 (15.4)
Finished primary	325 (36.0)
Finished Grade 10	266 (29.4)
Finished Grade 12	108 (11.9)
Degree/Diploma	12 (1.3)
Other	54 (6.0)
Enough Money to meet own needs	**N (%)**
None	354 (39.3)
A little	318 (35.3)
Moderately	112 (12.4)
Mostly	40 (4.4)
Completely	10 (1.1)
Other	67 (7.4)

Table 2. Demographic Characteristics.

Items	M (SD)
Mean no. of clinic visits in 12 months	8.48 (6.798)
Mean no. of hospital visits in 12 months	1.14 (1.905)
Mean no. of private doctor visits in 12 months	1.58 (2.520)
Mean no. of traditional healer visits in 12 months	0.23 (1.121)
Main reason for visiting health facility	N (%)
Non-communicable diseases	108 (11.5)
Communicable diseases	**40 (4.3)**
Treatment	**391 (41.8)**
Treatment (for baby or child)	**57 (6.1)**
Bodily aches	**47 (5.0)**
Family Planning	**95 (10.1)**
Other/Unidentified	**198 (21.2)**

Table 3. Health Care Utilization.

3.3 Symptom reporting
More than two-thirds reported that coughing, headache, fever, and body/limb aches were the symptoms suffered in descending order.

Symptoms	N (%)
Coughing	183 (87.1)
Body/limb aches	110 (80.3)
Fever	153 (85.0)
Rash	46 (63.0)
Headache	178 (86.8)
Diarrhoea	23 (46.0)

Table 4. Symptom Reporting.

3.4 Prior diagnosis
More than 60% of the respondents indicated that they had prior diagnosis of other STIs (95%), other illnesses (88.9%), TB (78.4%), Diabetes (72.2%), High Blood Pressure (69.3%) and HIV (65.2%).

Prior Diagnosis	N (%)
TB	105 (78.4)
HIV	56 (65.1)
Diabetes	78 (72.2)
Other STI	891 (94.9)
High Blood Pressure	651 (69.3)
Pregnancy	38 (55.1)
Other illness	835 (88.9)

Table 5: Prior diagnosis.

4. Descriptions of patients evaluations: percentage of patients who used the most positive answering category by sex (N=836, percentages)

4.1 Access to PHC services
A larger portion of women respondents positively agreed with the items from the access domain, than did men. More than three quarters of women agreed that it was possible to get an appointment that suited them and about the same number indicated that no payment was required for treatment at that clinic. About three quarters of women also agreed that it was possible to get through to the clinic by telephone and that the clinic was disability-friendly. The same number of women also agreed that they were treated by nurses who spoke a language they could understand and that the clinic's opening hours were convenient. Nearly 75% of women further agreed that they are always treated and not asked to return on another day and that is was possible to speak to the nurse on the phone. More than seven in ten women did not think that nurses did not visit their places of residence often enough. In terms of time and financial costs, about seven in ten women respondents agreed that the journey there took longer than one hour and that it costs more than R10-00 (US$ 1.46) to get to the clinic.

Item	Men: N (%)	Women: N (%)	Total: N (%)	P
It takes longer than an hour to go to the clinic	70 (27.5)	184 (72.2)	255 (27.2)	0.067
It cost more than R10.00 to get to the clinic	50 (29.2)	120 (70.2)	171 (18.2)	0.218
The clinic has convenient opening hours	161 (25.8)	461 (74.0)	623 (66.3)	0.982
I don't think healthworkers/nurses come often enough to the place where I stay	76 (27.4)	201 (72.6)	277 (29.5)	0.689
I paid money to be treated in this clinic	28 (24.6)	86 (75.4)	144 (12.1)	0.088
The nurse who treated me spoke in a language I understood	191 (25.5)	556 (74.3)	748 (79.7)	0.07
When I come to this clinic I'm always treated & never told to return on another day	155 (25.8)	445 (74.0)	601 (64.0)	0.478
The clinic is user friendly to disabled persons	153 (25.4)	449 (74.6)	602 (64.1)	0.265
Getting an appointment to suit you	114 (23.7)	367 (76.3)	481 (51.2)	0.429
Getting through to the clinic on the phone	101 (25.0)	303 (75.0)	404 (43.1)	0.834
Being able to speak to the nurse practitioner on the telephone	93 (26.6)	256 (73.4)	349 (37.2)	0.588

Table 6. Perceived Access to PHC Services.

4.2 Perceived empathy

Women, when compared to men, were also more positive in their responses to items of the empathy domain. More than three quarters of women respondents agreed that their privacy was respected by all the staff involved in their treatment, that the nurse/doctor who treated them was polite and that they could answer all questions about their illness. The same number felt that this made it easy to tell the doctor/nurse about their problems. Just under three quarters of women respondents agreed that the nurse/doctor who treated them introduced themselves, that they gave their permission to be examined and treated and made them feel they had time during consultations.

4.3 General satisfaction

Larger proportions of women when compared to men had positive responses on items of this domain. Almost eight in ten women respondents positively agreed that patients do not usually appreciate all that the clinic staff does for them. More than three in four also agreed that staff do inform clients of changes in service, as well as any delays in services, on occasion. The same proportion of women agreed that their treatment is always better when an injection is administered and that they are pleased with the way they were treated at the clinic. Nearly three quarters agreed that they always get treatment when attending the clinic where they were interviewed and that they would attend the same facility again on another occasion. The same number will also recommend the clinic to friends and family when should they be sick. Despite these figures, just more than six in ten women agreed that the staff were helpful.

Item	Men: N (%)	Women: N (%)	Total: N (%)	P
The nurse/Doctor who treated me introduced him/herself	121 (25.9)	348 (73.9)	468 (49.8)	0.49
The nurse/Doctor who treated me answered all questions about my illness	162 (24.8)	490 (75.0)	653 (69.5)	0.474
I gave permission to be examined and treated	185 (25.4)	543 (74.5)	729 (77.6)	0.908
My privacy was respected by all the staff	172 (24.2)	538 (75.7)	711 (75.7)	0.022
The nurse/doctor who treated me was polite	162 (24.3)	503 (75.5)	666 (70.9)	0.251
The nurse in this clinic are very interested in their clients	165 (25.2)	490 (74.7)	656 (69.9)	0.883
Making you feel you had time during consultations	168 (26.4)	469 (73.6)	638 (67.9)	0.481
Interest in your personal situation	163 (26.1)	462 (74.0)	626 (66.6)	0.425
Making it easy for you to tell him or her about your problems	164 (24.8)	497 (75.2)	662 (70.5)	0.721

Table 7. Empathy.

Item	Men: N (%)	Women: N (%)	Total: N (%)	P
Next time I am ill I will come back here	184 (25.0)	551 (74.9)	736 (78.4)	0.034
I was pleased with the way I was treated at this clinic	172 (24.0)	544 (75.9)	717 (76.4)	0.034
If my friends/family are sick I will tell them to come to this facility	172 (25.0)	514 (74.8)	687 (73.2)	0.426
My treatment is always better if I have an injection	117 (23.6)	379 (76.4)	496 (52.8)	0.147
Patients don't usually appreciate all that staff in this clinic do for them	97 (21.6)	353 (78.4)	450 (47.9)	0.008
I always get treatment when I come here	172 (24.6)	526 (75.3)	699 (74.4)	0.397
Staff informs clients of delays in service from time to time	136 (23.7)	438 (76.3)	574 (61.1)	0.042
Staff informs clients of changes in service from time to time	128 (23.3)	421 (76.7)	549 (58.5)	0.019
The helpfulness of staff	277 (37.3)	465 (62.7)	624 (66.4)	0.715

Table 8. General Satisfaction.

4.4 Referral

The items in this domain also received a majority of positive responses from women respondents. For example, more than three quarters of women agreed that if they cannot be helped at the clinic they will be referred to the nearest hospital or doctor. They same number was also sure that nurses in this facility will call an ambulance if a client is very sick and that

nurses in that facility ask patients to return to see how they are doing. The role of
traditional healers is still an important aspect of health care, as more than three quarters of
women responded that they usually visit a traditional healer before coming to the clinic.

Item	Men: N (%)	Women: N (%)	Total: N (%)	P
If I can't be helped here I will be referred to the nearest hospital/Doctor	163 (24.4)	503 (75.4)	667 (71.0)	0.216
Nurses in this facility call an ambulance if a client is very sick	167 (24.4)	517 (75.5)	685 (72.9)	0.358
Nurses in this facility ask patients to return to see how they are doing	152 (24.2)	475 (75.6)	628 (66.9)	0.321
When I'm sick I usually visit a traditional healer before I come to clinic	42 (23.3)	137 (76.1)	180 (19.2)	0.395

Table 9. Referral.

4.5 Service standards

Items in the service standards domain elicited more positive responses from women than
from men. More than three in four women responded that they knew either the chairperson
or a member of the clinic committee of that clinic that the health worker that assisted them
had a name tag on her/him, that they knew where and to whom to raise complaints, and
know of the availability of a suggestion box at the clinic. The same number also agreed that
the registration procedures in the clinic were satisfactory, waiting time before examination
was reasonable and that there were are fast queues in this clinic for certain services. Just
fewer than three in four women agreed that when they had reason to complain, they
received feedback and that such action improved service delivery.

Item	Men: N (%)	Women: N (%)	Total: N (%)	P
The registration procedures in this clinic are satisfactory	148 (24.7)	451 (75.2)	600 (63.9)	0.005
In this clinic the time I had to wait before I was examined was reasonable	127 (25.5)	372 (74.5)	499 (53.1)	0.449
There are fast queues in this clinic (e.g. under 5 Immunisation, TB clients, etc)	114 (24.9)	342 (74.8)	457 (48.7)	0.849
The health worker that assisted me had a name tag on him/her	150 (23.9)	476 (75.9)	627 (66.8)	0.248
I know where and to whom to raise my complaints	89 (26.0)	253 (74.0)	342 (36.4)	0.602
When I complain I write it and put it in the suggestion box provided	85 (24.7)	259 (75.3)	344 (36.6)	0.606
When I complained I received feedback	61 (24.6)	187 (75.4)	248 (26.4)	0.723
Raising complaints/suggestions improve service delivery	87 (27.0)	235 (73.0)	322 (34.3)	0.042
I know the chairperson/member of the clinic committee of this facility	48 (23.4)	157 (76.6)	205 (21.8)	0.515

Table 10. Service Standards.

4.6 Reliability

More than three quarters of women judged services as reliable as they did not wait long before receiving medication and that the clinic provided quick services for urgent health problems. Just under this figure regarded general waiting time in waiting rooms as positive.

Item	Men: N (%)	Women: N (%)	Total: N (%)	P
If I received medicines or pills I did not have to wait long for them	140 (24.2)	439 (75.8)	579 (61.7)	0.003
Waiting time in the waiting room	106 (25.2)	314 (74.8)	420 (44.7)	0.062
Providing quick services for urgent health problems	136 (24.2)	425 (75.8)	562 (59.8)	0.161

Table 11. Reliability.

4.7 Health promotion

A majority of women respondents were positive on items referring to health promotion at the clinic. For example, more than three quarters agreed that as patients are waiting to be seen, health workers in the clinic sometimes give talks on health related issues affecting the community. Also, nearly three in four replied that when they had to wait at the clinic, very useful things can be learnt from the posters and other IEC materials. The reason for this was that the posters and other IEC materials, the 'Batho Pele' (people first) principles and the patients' rights charter, were all in a language they could understand.

Item	Men: N (%)	Women: N (%)	Total: N (%)	P
I saw on the walls of this clinic a Patients Rights Charter in a language I could understand	148 (26.3)	413 (73.5)	562 (59.9)	0.887
I saw on the walls of this clinic Batho Pele Principles in a language I could understand	141 (26.3)	395 (73.6)	537 (57.2)	0.719
When I had to wait in this clinic I sometimes learn very useful things from the posters and other IEC (Information, Education & Communication) materials	138 (24.9)	415 (74.9)	554 (59.0)	0.417
The posters and other IEC material are in a language I understand	145 (25.0)	435 (74.9)	581 (61.9)	0.927
As patients are waiting to be seen, health workers in this facility sometimes talk to us about health related issues that affect our community	113 (22.9)	379 (76.9)	493 (52.5)	0.177

Table 12. Health Promotion.

4.8 Tangibles

Items under the tangibles domain also yielded positive responses from the majority of women respondents. More than three in four women agreed that the toilets were clean and in a good condition, that there were indeed toilets for patients in the clinic and that the clinic has enough consultation rooms and that there were enough benches for patients to sit while waiting to be seen by health workers. Just under three quarters agreed that there was clean drinking water for patients, that the building was in a good condition and the clinic and its surroundings are clean, that the services and hours of service displayed on the board outside the clinic was clear and in a language that could be understood.

Domain	Men: N (%)	Women: N (%)	Total: N (%)	P
The clinic building is in a good condition	174 (25.7)	501 (74.1)	676 (72.0)	0.976
The clinic and its surroundings are clean	182 (25.5)	532 (74.4)	715 (76.1)	0.903
There are toilets for patients in this clinic	167 (24.1)	526 (75.8)	694 (73.9)	0.026
The toilets are in a good condition	150 (23.7)	483 (76.2)	634 (67.5)	0
The toilets are clean	146 (23.7)	468 (76.1)	615 (65.5)	0.009
The clinic has enough consultation rooms	137 (24.5)	421 (75.3)	559 (59.5)	0.282
There are benches for patients to sit while waiting to be seen by health worker	165 (24.5)	507 (75.3)	673 (71.7)	0.615
There is clean water for patients in this clinic	171 (25.1)	509 (74.7)	681 (72.5)	0.432
The services rendered and hours of service are clearly displayed on a board outside the facility	149 (26.0)	424 (73.9)	574 (61.1)	0.463
The services and hours of service displayed on the board outside are in a language I can understand	142 (25.2)	421 (74.6)	564 (60.1)	0.378

Table 13. Tangibles.

4.9 Assurance

A greater percentage of women than men also responded positively to items of the assurance domain. More than three quarters of women agreed that the staff at the clinic had given preferential treatment to patients who looked more ill, that the nurses were able to tell them more about their illness and symptoms, and that were told how to store and self-administer their medication. The same proportion also agreed that health workers gave them help in dealing with the emotional problems related to your health status, that they felt comfortable to bring their partners to the facility when requested and that they felt assured that their treatment records remained confidential. Exactly three quarters agreed with their physical examination by health workers, and that they helped in making their patients understand the importance of following his or her medical advice and their preparation of patients as to what to expect from specialist or hospital care.

Just under three quarters of women respondents agreed that attending the health service meant quick relief of one's symptoms, that the explanation of the purpose of tests and treatments were clear and that they felt compelled to complete their treatment as was instructed. Slightly less than three in four women agreed that health workers at the facility listened to patients, that they can get them to always return when asked to do so and that they involved their patients in decisions affecting their medical care. The same proportion of women felt that health workers also helped patients to feel well enough to perform normal daily activities, were through, knew what advice were given to patients previously and were competent in offering advice on the prevention of diseases.

Domain	Men: N (%)	Women: N (%)	Total: N (%)	P
At the time I was waiting to be seen by a Health Worker there was a patient that looked more ill	89 (23.8)	285 (76.2)	374 (39.8)	0.172
I always return when asked by the nurse to come back	183 (25.4)	536 (74.4)	720 (76.7)	0.381
I finish all my treatment as instructed	190 (25.3)	559 (74.5)	750 (79.9)	0.885
I bring my partner(s) when requested to	157 (24.7)	478 (75.2)	636 (67.7)	0.852
I was told how to take my pills/medication	187 (24.6)	572 (75.3)	799 (80.9)	0.433
I was told how to store my pills/medication	171 (24.5)	527 (75.4)	699 (74.4)	0.136
Involving you in decisions about your medical care	168 (25.8)	483 (74.2)	652 (69.4)	0.608
Listening to you	182 (25.5)	533 (74.5)	716 (76.3)	0.934
Keeping your records and data confidential	175 (24.6)	529 (75.1)	705 (75.0)	0.956
Quick relief of your symptoms	167 (25.4)	490 (74.6)	658 (70.1	0.917
Helping you to feel well so that you can perform your normal daily activities	172 (25.9)	493 (74.1)	665 (70.9)	0.213
Thoroughness	150 (26.5)	416 (73.5)	567 (60.4)	0.499
Physical examination of you	155 (25.0)	466 (75.0)	622 (66.3)	0.202
Offering you services for preventing diseases	168 (26.2)	473 (73.8)	642 (68.4)	0.228
Explaining the purpose of tests and treatments	173 (25.4)	509 (74.6)	683 (72.7)	0.356
Telling you what you wanted to know about your symptoms and/or illness	162 (24.0)	514 (76.0)	678 (72.2)	0.604
Help in dealing with emotional problems related to your health status	161 (24.7)	490 (75.3)	651 (69.3)	0.237
Helping you understand the importance of following his or her advice	168 (25.1)	502 (75.0)	671 (71.4)	0.488
Knowing what s/he had done or told you during previous contacts	162 (26.0)	461 (74.0)	624 (66.5)	0.406
Preparing you for what to expect from specialist or hospital care	155 (25.1)	463 (75.0)	619 (66.0)	0.914

Table 14. Assurance.

5. Discussion

Seeking to understand patient perspectives is an important step in the efforts to improve the quality of health care. Research examining patient satisfaction with health care provision in South Africa and, more specifically, the perceived quality of care given by the health care providers is limited (Myburgh et al., 2005). In this study, there were consistently significant differences regarding patient satisfaction between male and female patients across selected items in the various domains.

Evidence from developed countries for gender differences in mean satisfaction levels is mixed. Some authors report that women are more satisfied than men with medical care received (Weiss, 1988), and some report that women are more critical of medical care than

Patient Satisfaction with Primary Health Care Services in a Selected District Municipality of
the Eastern Cape of South Africa

71

men #(Kaplan, 1996), whilst a 2005 Canadian study (Human Resources and Skills Development Canada, 2009) found almost similar satsifaction levels between male (86%) and female (84%) patients. However, a meta-analysis of 110 studies of patient satisfaction, using standard instruments, concluded that there was no average difference in satisfaction with medical care between women and men (Hall & Dornan, 1990). More recently, Sanmartin et al. (2002) suggested that user frequency might influence the descrepancies found betwee male and female patient statisfaction rates and that the type of service being assessed might be a further factor.

Wessels et al (2010) found that amongst oncology patients women rated care aspect of services more highly. A recent Ugandan study found some gender and age difference in patient satisfaction with TB services (Babikako et al. 2011). Past experience and consequently patient expectations, they argue might influence age and gender differences in patient satisfaction.

What is common across these studies is the importance of considering the influence of demographic variables on patient satisfaction. Simply controlling for demographic differences, might result in the needs of important demographic groupings being overlooked. In addition, demographic differences, such as gender are likely to shape patients' needs and preferences and might be a particularly important consideration in shaping specific health services to better meet needs and support treatment adherence. In the South African context, the role that gender plays in patient satisfaction and the gender differences in patient satisfaction need further exploration. We conclude that quality improvement and research in primary care could benefit from gender analysis of patient satisfaction data and from more gender-sensitive patient satisfaction measures.

6. Acknowledgements

We thank the Eastern Cape Department of Health for financially supporting the study.
We would like to extend our gratitude to the district manager and clinic managers of the 12 clinics for overseeing the study in their respective clinics.
Our gratitude is also due to the fieldworkers and patients who agreed to be interviewed.

7. References

Ajayi, I.O., Olumide, E.A. & Oyediran, O., 2005, 'Patient satisfaction with the services provided at a general outpatients' clinic, Ibadan, Oyo State, Nigeria', *African Journal of Medicine & Medical Science* 34(2), 133-140.

Andaleeb, S.S., 2001, 'Service quality perceptions and patient satisfaction: a study of hospitals in a developing country', *Social Science & Medicine* 52, 1359–1370.

Babikako, H.M., Neuhauser, D., Katamba, A., Mupere, E.(2011). Patient satisfaction, feasibility and reliability of satisfaction questionnaire among patients with pulmonary tuberculosis in urban Uganda: a cross-sectional study. *Health Research Policy and Systems* 2011, 9:6. Available from:
http://www.health-policy-systems.com/content/9/1/6

Bediako, M.A., Nel, M. & Hiemstra, L.A., 2006, 'Patients' satisfaction with government health care and services in the Taung district, North West Province' *Curationis* 29(2), 12-15.

Bradshaw D, Groenewald P, Laubscher R, Nannan N, Nojilana B, Norman R, Pieterse D and Schneider M. *Initial Burden of Disease Estimates for South Africa, 2000.* Cape Town: South African Medical Research Council, 2003.

Brislin, R.W., 1970, 'Back translation for cross-cultural research', *Journal of Cross-Cultural Psychology* 1(3), 185-216.

Campbell, J.L., Ramsay, J. & Green, J., 2001, 'Age, gender, socioeconomic and ethnic differences in patients' assessments of primary health care', *Quality in Health Care* 10, 90-95.

Dağdeviren, N. & Akturk, Z., 2004, 'An evaluation of patient satisfaction in Turkey with the EUROPEP instrument', *Yonsei Medical Journal* 45(1), 23-28.

De Jager ,J., & Du Plooy, T., 2007, 'Service quality assurance and tangibility for public health care in South Africa' *Acta Commercii* 7, 96-117.

Department of Health, 2007, *A policy on quality in health care for South Africa*, Department of Health, Pretoria.

Department of Public Service and Administration, 1997, *Transforming Public Service Delivery*, Department of Public Service and Administration, Pretoria.

Eastern Cape Socio-Economic Consultative Council (ECSECC) 2011. Statistics at your fingertips *http://www.ecsecc.org/statistics-database* [Accessed 28 June 2011]

Glick, P., 2009, 'How reliable are surveys of client satisfaction with healthcare services? Evidence from matched facility and household data in Madagascar', *Social Science & Medicine* 68(2), 368-379.

Hall, J.A. & Dornan, M.C., 1990, 'Patient socio-demographic characteristics as predictors of satisfaction with medical care: A meta-analysis, *Social Science & Medicine*, 30(7), 811-818.

Hospital Association of South Africa, 2011. *http://www.hasa.co.za/hospitals/members/* [Accessed 28 June 2011]

Kaplan SH, Sullivan LM, Spetter D. *Gender and patterns of physician-patient communication.* In: Falik MM, Collins KS, eds. Women's health: The Commonwealth Fund Survey. Baltimore, MD: Johns Hopkins University Press; 1996.

McIntyre D, 2010, *Private sector involvement in funding and providing health services in South Africa: implications for equity and access to health care*, EQUINET Discussion Paper Series 84 Health Economics Unit (UCT), ISER Rhodes University, EQUINET: Harare.

Muhondwa, E.P., Leshabari, M.T., Mwangu, M., Mbembati, N. & Ezekiel, M.J., 2008, 'Patient satisfaction at the Muhimbili National Hospital in Dar es Salaam, Tanzania', *East African Journal of Public Health* 5(2), 67-73.

Myburgh, N.G., Solanki, G.C., Smith, M.J. & Lalloo, R., 2005, 'Patient satisfaction with health care providers in South Africa: the influences of race and socioeconomic status', *International Journal for Quality in Health Care* 17(6), 473-477.

Peltzer, K., 2009, 'Patient experiences and health system responsiveness in South Africa' , *BMC Health Services Research* 9, 117, DOI:10.1186/1472-6963-9-117.

Peltzer, K., 2000, 'Community perceptions of biomedical health care in a rural area in the Limpopo Province South Africa', *Health SA Gesondheid* 5 (1), 55-63.

Newman, R.D., Gloyd, S., Nyangezi, J.M., Machobo, F. & Muiser, J., 1998, 'Satisfaction with Outpatient Health Care Services in Manica Province, Mozambique', *Health Policy & Planning* 13(2), 174-180.

Sanmartin, C., Houle, C., Berthelot, J. & White, K., 2002, *Access to Health Care Services in Canada, 2001;* Statistics Canada, Ottawa.

Weiss, G.L., 1988, 'Patient Satisfaction with Primary Medical Care Evaluation of Sociodemographic and Predispositional Factors', *Medical Care* 26(4), 383-392.

Wessels, H., De Graff, A., Wynia, K., De Heus, M., Kruitwagen, C.L.J.J., Woltjer, G.T.G.J., Teunissen, S.C.C.M., Voest, E. (2010). Gender-Related Needs and Preferences in Cancer Care Indicate the Need for an Individualized Approach to Cancer Patients. *The Oncologist,*15:648–655 doi: 10.1634/theoncologist.2009-0337

5

Analysis of MLPA Data Using Novel Software Coffalyser.NET by MRC-Holland

Jordy Coffa[1,2] and Joost van den Berg[3]
[1]MRC-Holland
[2]Free University Amsterdam
[3]Berg IT solutions
The Netherlands

1. Introduction

Genetic knowledge has increased tremendously in the last years, filling gaps and giving answers that were inaccessible before. Medical genetics seeks to understand how genetic variation relates to human health and disease (National Center for Biotechnology Information, 2008). Although genetics plays a larger role in general, the knowledge of the genetic origins of disease has increased our understanding of illnesses caused by abnormalities in the genes or chromosomes, offering the potential to improve the diagnosis and treatment of patients. Normally, every person carries two copies of every gene (with the exception of genes related to sex-linked traits), which cells can translate into a functional protein. The presence of mutant forms of genes (mutations, copy number changes, insertion/deletions and chromosomal alterations) may affect several processes concerning the production of these proteins often resulting in the development of genetic disorders. Genetic disease is either caused by changes in the DNA of somatic cells in the body or it is inherited, e.g. by mutations in the germ cells of the parents.

Genetic testing is "the analysis of, chromosomes (DNA), proteins, and certain metabolites in order to detect heritable disease-related genotypes, mutations, phenotypes, or karyotypes for clinical purposes (Holtzman et al, 2002). In order to make this suitable for routine diagnostics dedicated, affordable, fast, easy-to-interpret and simple-to-use genetic tests are necessary. This allows scientists to easily access information that for instance can be used to: confirm or rule out a suspected genetic condition or help determine a person's chance of developing or passing on a genetic disorder. Several hundred genetic tests are currently in use, and more are being developed (Sequeiros et al, 2008). The Multiplex Ligation-dependent Probe Amplification (MLPA) is a PCR-based technique, which allows the detecting of copy number changes in DNA or RNA. MLPA can quantify up to 50 nucleic acid sequences or genes in one simple reaction, with a resolution down to the single nucleotide level (Schouten et al., 2002) needing only 20 ng of DNA. The MLPA procedure itself needs little hands on work allowing up to 96 samples to be handled simultaneously while results can be obtained within 24 hours. These properties make it a very efficient technique for medium-throughput screening of many different diseases in both a research and diagnostic settings (Ahn et al, 2007).

Over a million of MLPA reactions were performed last year worldwide but researchers are still concerned with the application of tools to facilitate and improve MLPA data analysis on large, complex data sets. MLPA kits contain oligo-nucleotide probes that through a biochemical reaction can produce signals that are proportional to the amount of the target sequences present in a sample. These signals are detected and quantified on a capillary electrophoresis device producing a fragment profile. The signals of an unknown sample need to be compared to a reference in order to assess the copy number. Profile comparison is a matter of professional judgment and expertise. Diverse effects may furthermore systematically bias the probe measurements such as: quality of DNA extraction, PCR efficiency, label incorporation, exposure, scanning, spot detection, etc., making data analysis even more challenging. To make data more intelligible, the detected probe measurements of different samples need to be normalized thereby removing the systematic effects and bringing data of different samples onto a common scale.

Although several normalization methods have been proposed, they frequently fail to take into account the variability of systematic error within and between MLPA experiments. Each MLPA study is different in design, scope, number of replicates and technical considerations. Data normalization is therefore often context dependent and a general method that provides reliable results in all situations is hard to define. The most used normalization strategy therefore remains the use of in-house brew analysis spreadsheets that often cannot provide the reliability required for results with clinical purposes. These sheets furthermore do not provide easy handling of large amounts of data and file retrieval, storage and archival needs to be handled by simple file management systems. We therefore set out to develop software that could tackle all of these problems, and provide users with reliable results that are easy to interpreter.

In this chapter we show the features and integrated analysis methods of our novel MLPA analysis software called Coffalyser.NET. Our software uses an analysis strategy that can adapt to fit the researcher objectives while considering both the biological context and the technical limitations of the overall study. We use statistical parameters appropriate to the situation, and apply the most robust normalization method based on the biology and quality of the data. Most information required for the analysis is extracted directly from the MRC-Holland database, producer of the MLPA technology, needing only little user input about the experimental design to define an optimal analysis strategy. In the next section we review the MLPA technology in more detail and explain the principles of MLPA data normalization. Then in section 3, we describe the main features of our software and their significance. The database behind our software is reviewed in section 4 and section 5 explains the exact workflow of our program reviewing the importance and methodology of each analysis step in detail. In the final section, we summarize our paper and present the future directions of our research.

2. Background

MLPA data is commonly used for sophisticated genomic studies and research to develop clinically validated molecular diagnostic tests, which e.g. can provide individualized information on response to certain types of therapy and the likelihood of disease recurrence. The most common application for MLPA is the detection of small genomic aberrations, often accounting for 10 to 30% of all disease-causing mutations (Redeker et al., 2008). In case of the very long DMD gene –involved in Duchenne Muscular Dystrophy – exon deletions and duplications even account for 65-70% of all mutations (Janssen et al., 2005). Since MLPA can detect sequences that differ only a single nucleotide, the technique is also widely used for the analysis of complicated diseases such as congenital adrenal hyperplasia and spinal muscular atrophy, where pseudo-genes and gene conversion complicate the analysis (Huang et al., 2007). Methylation-specific MLPA has also proven to be a very useful method for the detection of aberrant methylation patterns in imprinted regions such as can be found with the Prader-Willi/Angelman syndrome and Beckwith-Wiedemann syndrome (Scott et al., 2008). The MS-MLPA method can also be used for the analysis of aberrant methylation of CpG islands in tumour samples using e.g. DNA derived from formalin-fixed, paraffin-embedded tissues.

MLPA kits generally contain about 40-50 oligo-nucleotide probes targeted to mainly the exonic regions of a single or multiple genes. The number of genes that each kit contains is dependent on the purpose of the designed kit. Each oligo-probe consists of two hemi-probes, which after denaturation of the sample DNA hybridize to adjacent sites of the target sequence during an overnight incubation. For each probe oligo-nucleotide in a MLPA kit there are about 600.000.000 copies present during the overnight incubation. An average MLPA reaction contains 60 ng of human DNA sample, which correlates to about 20.000 haploid genomes. This abundance of probes as compared to the sample DNA allows all target sequences in the sample to be covered. After the overnight hybridization adjacent hybridized hemi-probe oligo-nucleotides are then ligated using a ligase enzyme and the ligase cofactor NAD at a slightly lower temperature than the hybridization reaction (54 °C instead of 60 °C). The ligase enzyme used, Ligase-65, is heat-inactivated after the ligation reaction. Afterwards the non-ligated probe oligonucleotides do not have to be removed since the ionic conditions during the ligation reaction resemble those of an ordinary 1x PCR buffer. The PCR reaction can therefore be started directly after the ligation reaction by adding the PCR primers, polymerase and dNTPs. All ligated probes have identical end sequences, permitting simultaneous PCR amplification using only one primer pair. In the PCR reaction, one of the two primers is fluorescently labeled, enabling the detection and quantification of the probe products.

The different length of every probe in the MLPA kit then allows these products to be separated and measured using standard capillary fragment electrophoresis. The unique length of every probe in the probe mix is used to associate the detected signals back to the original probe sequences. These probe product measurements are proportional to the amount of the target sequences present in a sample but cannot simply be translated to copy numbers or methylation percentages. To make the data intelligible, data of a probe originating from an unknown sample needs to be compared with a reference sample. This reference sample is usually performed on a sample that has a normal (diploid) DNA copy number for all target sequences. In case the signal strengths of the probes are compared with those obtained from a reference DNA sample known to have two copies of the chromosome, the signals are expected to be 1.5 times the intensities of the respective probes from the reference if an extra copy is present. If only one copy is present the proportion is expected to be 0.5. If the sample has two copies, the relative probe strengths are expected to be equal. In some circumstances reliable results can be obtained by comparing unknown samples can to reference samples by visual assessment, simply by overlaying two fragment profiles and comparing relative intensities of fragments (figure 1).

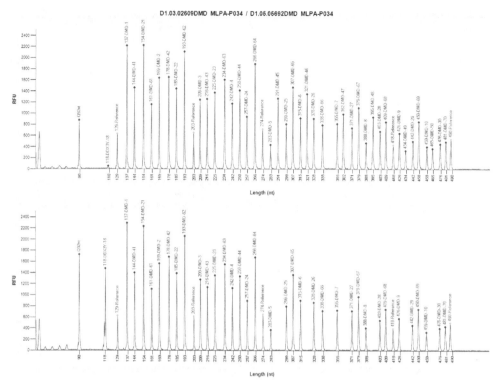

Fig. 1. MLPA fragment profile of a patient sample with Duchenne disease (bottom) and that of a reference sample (top). Duchenne muscular dystrophy is the result of a defect in the DMD gene on chromosome Xp21. The fragment profile shows that the probe signals targeted to exon 45-50 of the DMD gene have a 100% decrease as compared to the reference, which may be the result of a homozygous deletion.

It may however not be feasible to obtain reliable results out of such a visual comparison if:
1. The DNA quality of the samples and references is incomparable.
2. The MLPA kit contains probes targeted to a number of different genes or different chromosomal regions, resulting in complex fragment profiles
3. The data set is very large, making visual assessment very laborious.
4. The DNA was isolated tumor tissue, which often shows DNA profiles with altered reference probes

To make (complex) MLPA data easier understandable unknown and reference samples have to be brought on a common scale. This can be done by normalization, the division of multiple sets of data by a common variable in order to cancel out that variable's effect on the data. In MLPA kits, so called reference probes are usually added, which may be used in multiple ways in order to comprise a common variable. Reference probes are usually are targeted to chromosomal regions that are assumed to remain normal (diploid) in DNA of applicable samples. The results of data normalization are probe ratios, which display the balance of the measured signal intensities between sample and reference. In most MLPA studies, comparing the calculated MLPA probe ratios to a set of arbitrary borders is used to recognize gains and losses (González, 2008). Probe ratios of below 0.7 or above 1.3 are for instance regarded as indicative of a heterozygous deletion (copy number change from two to one) or duplication (copy number change from two to three), respectively. A delta value of 0.3 is a commonly accepted empirically derived threshold value for genetic dosage quotient analysis (Bunyan et al. 2004). To get more conclusive results probes may be arranged according to chromosomal location as this may reveal more subtle changes such as those observed in mosaic cases.

3. Key features

3.1 Support wide range of file format

Our software is compatible with binary data files produced by all major capillary electrophoresis systems including: ABIF files (*.FSA, *.AB1, *.ABI) produced by Applied Biosystems devices, SCF and RSD files produced by MegaBACE™ systems (Amersham) and SCF and ESD files produced by CEQ systems (Beckman). We can also import fragment lists in text or comma separate format, produced by different fragment analysis software programs such as Genescan (Applied Biosystems), Genemapper (Applied Biosystems), CEQ Fragment analysis software (Beckman) and Genetools. Raw data files are however preferred since they allow more troubleshooting and quality check options as compared to size called fragment lists. Next to this, raw and analyzed data are then stored in a single database and more advanced reports can be created.

3.2 Optimized peak detection / quantification method for MLPA

All applied algorithms in our software are specifically designed to suit MLPA or MLPA-like applications. We designed an algorithm for peak detection and quantification specifically for MLPA peak patterns. Most peak detection algorithms simply identify peaks based on amplitude, ignoring the additional information in the shape of the peaks. In our experience, 'true' peaks have characteristic shapes, and including fluorescence of artifacts may introduce ambiguity into the analysis and interpretation process. Our algorithm has the ability to differentiate most spurious peaks and artifacts from peaks that originate from a probe product. We differentiate a number of different peak artifacts, such as: shoulder peaks, printout spikes, dye artifacts, split peaks, pull-up peaks, stutter peaks and non-template additions. It is often difficult to identify the correct peaks due to appearance of nonspecific peaks in the vicinity of the main allele peak. Our algorithm is therefore optimized to discriminate the different artifacts from the probe signals by usage of minimum and maximum threshold values on the peak -amplitude, -area, -width and -length. Next to this, it may also recognize split and shoulder -peaks by means of shape recognition, making correct identification of probe signals even more reliable. Following peak detection, quantification and size calling, our software allows one or more peaks to be linked to the original MLPA probe target sequence. This pattern matching is greatly simplified as compared to other genotyping programs and additionally provides a powerful technique for identifying and separating signal due to capillary electrophoresis artifacts. Our software may employ three different metrics to reflect the amount of probe fluorescence: peak height, peak area and peak area including its siblings. Peak siblings are the peak artifacts that are created during the amplification of the true MLPA products but have received an alternative length. To determine which metric should be used for data normalization, our program uses an algorithm that compares the signal level of each metric over the reference probes in all samples, and compares this to the amount of noise over the same signals. The metric that has the largest level signal to noise is then used in the following normalization steps.

3.3 Performances and throughput

After a user logs in, analysis of a complete experiment can be performed in two simple steps: the processing of raw data and the comparison of different samples. Depending on the analysis setup and type of computer, the complete analysis may be completed in less than a minute for 24 samples. Our software can also make use of extra cores running in a computer, multiplying the speed of the analysis almost by two for each core. Because of problems arising from poor sample preparations, presence of PCR artifacts, irregular stutter bands, and incomplete fragment separations, a typical MLPA project requires manual examination of almost all sample data. Our software was designed to eliminate this bottleneck by substantially minimizing the need to review data. By creating a series of quality scores to the different processes users can easily pinpoint the basis for the failed analysis. These scores include quality assessment related to: the sample DNA, MLPA reaction, capillary separation and normalization steps (figure 6). The quality of each step can fall roughly into three categories.

1. High-quality or green. The results of these analysis steps can be accepted without reviewing.
2. Low-quality or red. These steps represent samples with contamination and other failures, which render the resulted data unsuitable to continue with. This data can quickly be rejected without reviewing; recommendations can be reviewed in Coffalyser.NET and used for troubleshooting.
3. Intermediate-quality or yellow. The results of these steps fall between high- and low-quality. The related data and additional recommendations can be reviewed in Coffalyser.NET and used to optimize the obtained results.

When the analysis is finished the results can be visualized in a range of different display and reporting options designed to meet the requirement of modern research and diagnostic facilities. Results effortlessly can be exported to all commonly used medical report formats such as: pdf, xls, txt, csv, jpg, gif, png etc.

3.4 Reliable recognition of aberrant probes

Results interpretation of clinically relevant tests can be one of the most difficult aspects of MLPA analysis and is a matter of professional judgment and expertise. In practice, most users only consider the magnitude of a sample test probe ratio, comparing the ratio against a threshold value. This criterion alone may often not provide the conclusive results required for diagnosing disease. MLPA probes all have their own characteristics and the level of increase or decrease that a probe ratio displays that was targeted to a region that contains a heterozygous gain or loss, may differ for each probe. Interpretation of normalized data may even be more complicated due to shifts in ratios caused by sample-to-sample variation such as: dissimilarities in PCR efficiency and size to signal sloping. Other reasons for fluctuations in probe ratios may be: poor amplification, misinterpretation of an artifact peak/band as a true probe signal, incorrect interpretation of stutter patterns or artifact peaks, contamination, mislabeling or data entry errors (Bonin et al., 2004). To make result interpretation more reliable our software combines effect-size statistics and statistical interference allowing users to evaluate the magnitude of each probe ratio in combination with it's significance in the population. The significance of each ratio can be estimated by the quality of the performed normalization, which can be assessed two factors: the robustness of the normalization factor and the reproducibility of the sample reactions.

During the analysis our software estimates the reproducibility of each sample type in a performed experiment by calculating the standard deviation of each probe ratio in that sample type population. Since reference samples are assumed to be genetically equal, the effect of sample-to-sample variation on probe ratios of test probes is estimated by the reproducibility of these probes in the reference sample population. These calculations may be more accurate under circumstances where reference samples are randomly distributed across the performed experiment. Our program therefore provides an option to create a specific experimental setup following these criteria, thereby producing a worksheet for the wet analysis and a setup file for capillary electrophoresis devices. DNA sample names can be selected from the database and may be typed as a reference or test sample, positive control or negative control. This setup file replaces the need for filling in the sample names in the capillary electrophoresis run software thereby minimizing data entry errors.

To evaluate the robustness of the normalization factor our algorithm calculates the discrepancies computed between the probe ratios of the reference probes within each sample. Our normalization makes use of each reference probe for normalization of each test probe; thereby producing as many dosage quotients (DQ) as there are references probes. The median of these DQ's will then be used as the definite ratio. The median of absolute deviations between the computed dosage quotients may reflects the introduced mathematical imprecision of the used normalization factor. Next, our software calculates the effect of both types of variation on each test sample probe ratio and determines a 95% confidence range. By comparing each sample's test probe ratio and its 95% confidence range to the available data of each sample type population in the experiment, we can conclude if the found results are significantly different from e.g. the reference sample population or equal to a positive sample population. The algorithm then completes the analysis by evaluating these results in combination with the familiar set of arbitrary borders used to recognize gains and losses. A probe signal in concluded to be aberrant to the reference samples; if a probe signal is significantly different as from that reference sample populations and if the extent of this change meets certain criteria. The results are finally translated into easy to understand bar charts (figure 2) and sample reports allowing users to make a reliable and astute interpretation of the results.

3.5 Advanced data mining options

The database behind our software is designed in SQL and is based on a relational database management system (RDBMS). In short this means that data is stored in the form of tables and the relationship among the data is also stored in the form of tables. Our database setup contains a large number of subtraction levels, not only allowing users to efficiently store and review experimental sample data, but also allowing users to get integrative view on comprehensive data collections as well as supplying an integrated platform for comparative genomics and systems biology. While all data normalization occurs per experiment, experiments can be organized in projects, allowing advanced data-mining options enabling users to retrieve and review data in many different ways. Users can for instance review multiple MLPA sample runs from a single patient in a single report view. Results of multiple MLPA mixese may be clustered together, allowing users gain more confidence on any found results. The database can further handle an almost unlimited number of specimens for each patient, and each specimen can additionally handle an almost unlimited number of MLPA sample runs. To each specimen additional information can be related such as sample type, tissue type, DNA extraction method, and other clinical relevant data, which can be used for a wide range of data mining operations for discovery purposes. Some of these operations include:

1. Segmenting patients accurately into groups with similar health patterns.
2. Evidence based medicine, where the information extracted from the medical literature and the corresponding medical decisions are key information to leverage the decision made by the professional.

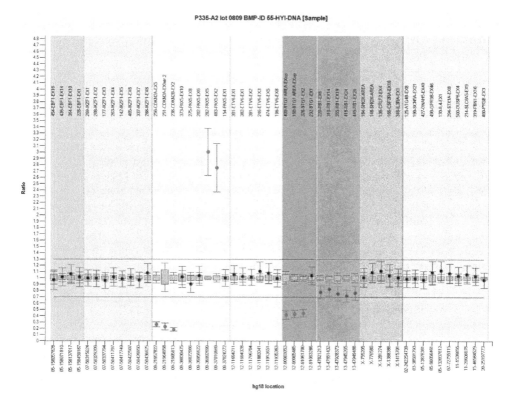

Fig. 2. Ratio chart of the results of a tumor sample analyzed with the P335 MLPA kit. Red dots display the probe ratios and the error bars the 95% confidence ranges. The orange box plots in the background show the 95% confidence range of the used reference samples. Map view locations are displayed on the x-axis and ratio results on the Y-axis. The red and green lines at ratio 0.7 and 1.3 indicate the arbitrary borders for loss and gain respectively. The displayed sample contains several aberrations and extra caution with interpretation is needed due to normal cell contamination.

3. Non-parametric tests (distribution-free) used to compare two or ore independent groups of data.
4. Classification methods that can be used for predicting medical diagnosis.

4. About the database

4.1 Client server database model

Our software uses a SQL client–server database model to store all project/experiment-related data. The client-server model has one main application (server) that deals with one or several slave applications (clients). Clients may communicate to a server over the network, allowing data sharing within and even beyond their institutions. Even though this system may provide great convenience e.g. for people who are working on a single project but are working on different locations, both client and server may also reside in the same system. Having both client and server on the same system has some advances over running both separately: the database is better protected and both client and server will always have the same version number. In case an older client will try to connect to a server that has a newer version number, the client needs to be updated first. A client does not share any of its resources, but requests a server's content or service function. Clients therefore initiate communication sessions with servers that await incoming requests. When a new client is installed on a computer it will implement a discovery protocol in order to search for a server by means of broadcasting. The server application will then answer with its dynamic address that resolves any issues with dynamic IP addresses.

4.2 User access

In addition to serving as a common data archive, the database provides user authentication, robust and scalable data management, and flexible archive capabilities via the utilities provided within Software. Our database model acts in accordance with a simple legal system, linking users to one or multiple organizations. Each user receives a certain role within each organization to which certain right are linked. These rights may for instance include denial of access to certain data but may also be used to deny access to certain parts of the program. These same levels may also be applied on project level. Projects will have project administrators and project members. The initial project creators will also be the project administrators who are responsible for user management of that project.

4.3 Sessions

As soon as a user makes a connection with the server a session will be started with a unique identifier. Subsequent made changes by any user will be held to this identifier, in order to keep track of the made changes. This number is also used to secure experiment data when in use; this ensures no two users try to edit essential data simultaneously (data concurrency). When a user logs in on a certain system, all previously open session of that user will be closed. Every user can thus only be active on a single system. On closing a session, either by logout or by double login all old user locks will disappear.

4.4 Data retrieval and updates

In our software is equipped with MLPA sheet manager software, allowing users to obtain information about commercial MLPA kits and size markers directly from the MRC-Holland database. Next to this, the sheet manager also allows users to create custom MLPA mixes.

The sheet manager software can be used to check if updates to any of the MLPA mixes are available. The sheet manager can further carry out automatic checks for updates at the frequency you choose, or it can be used to make manual checks whenever you wish. It can display scheduled update checks and can work completely in the background if you choose. With just one click, you can check to see if there are new versions of the program, or updated MLPA mix sheets. If updates are available, you can download them quickly and easily. In case some MLPA mixes are already in use, users may choose to hold on to both the older version and updated versions of the mix or replace the older version.

5. Coffalyser.NET workflow

Figure 3 shows the graphical representation of the workflow of our software. After creating an empty solution, users can add new or existing items to the empty solution by using the "add new project" or "add new experiment" command from the client software context menu. By creating projects, users can collect data of different experiments in one collection. Next, data files can then be imported to the database and linked to an experiment. Users then need to define for each used channel or dye stream of each capillary (sample run) what the contents are. Each detectable dye channel can be set as a sample (MLPA kit) or a size marker. Samples may further be typed as: MLPA test sample, MLPA reference sample, MLPA positive control, or MLPA digested sample. The complete analysis of each MLPA experiment can be divided in 2 steps: raw data analysis and comparative analysis. Raw data analysis includes all independent sample processes such as: the recognition and signal determination of peaks in the raw data streams of imported data files, the determination of the sizes of these peaks in nucleotides and the process of linking these peaks to their original probe target sequences. After raw data analysis is finished, users can evaluate a number of quality scores (figure 6), allowing users to easily assess the quality of the produced fragment data for each sample. Users may now reject, accept and adjust sample types before starting the comparative analysis. During the comparative part of the analysis several normalization and regression analysis methods are applied in order to isolate and correct the amount of variation that was introduced over the repeated measured data. Found variation that could not be normalized out of the equation is measured and used to define confidence ranges. The software finally calculates the variation of the probes over samples of the same types, allowing subsequent by classification of unknown samples. After the comparative analysis is finished, users may again evaluate a number of quality scores this time concerning the quality of different properties related to the normalization. The users can finally evaluate the results by means of reporting and visualization methods.

5.1 Import / export of capillary data
Importing data is the process of retrieving data from files to the SQL Server™ (for example, an ABIF file) and inserting it into SQL Server tables. Importing data from an external data source is likely to be the first step you perform after setting up your database. Our software contains several algorithms to decode binary files from the most commonly used capillary electrophoresis devices (see paragraph 2.1). Capillary devices usually store measurements of relative fluorescent units (RFU) and other related data that is collected during fragment separation in computer files encoded in binary form. Binary

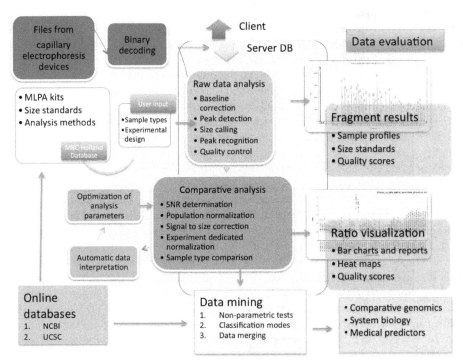

Fig. 3. Schematic overview of the Coffalyser.NET software workflow.

files are made up as a sequence of bytes, which our program decodes back into lists of the different measurements. The most important measurement being the laser induced fluorescence of the covalently bound fluorescent tags on the probe products and the size marker. The frequency at which these measurements occur depends on the type of system. A complete scan will always check all filters (or channels) and result in one data point. Almost all capillary systems are able to detect multicolor dyes permitting the usage of an internal size marker providing a more accurate size call than the usage of external size marker. Multicolor dyes may also permit the analysis of loci with overlapping size ranges, thus allowing multiple MLPA mixes to be run simultaneously in different dye colors.

After data has been imported into your SQL Server database, users can start the analysis. Users can choose to analyze the currently imported data or data that was imported in the past or a combination of both. Due to the relative nature of all MLPA data, it is recommended to analyze data within the confinements of each experiment. There do exist circumstances in which better results may be obtained by applying older collected reference data but one should use these options with caution.

Exporting data is usually a less frequent occurrence. Coffalyser.NET therefore does not have standard tools to export raw capillary data but rather depends on the provided tools and features of the SQL server. The data may be exported to a text file and then be read by third party applications such as Access or Microsoft Excel, which can then be used to view or manipulate the data.

5.2 Raw data analysis
5.2.1 Baseline correction

When performing detection of fluorescence in capillary electrophoresis devices it is some times the case that spectra can be contaminated by fluorescence. Baseline curvature and offset are generally caused by the sample itself and little can be designed in an instrument to avoid these interferences (Nancy T. Kawai, 2000). Non-specific fluorescence or background auto fluorescence should be subtracted from the fluorescence obtained from the probe products to obtain the relative fluorescence as a result of the incorporation of the fluorophore. The baseline wander of the fluorescence signals may cause problems in the detection of peaks and should be removed before starting peak detection. Our software corrects for this baseline by applying two times a median signal filter on the raw signals. First, the signals of the first 200 data points of each dye channel are extracted and its median was calculated. Then for every 200 subsequent data points till the end of the data stream, the same procedure is carried out. These median values are then subtracted from the signal of the original data stream to remove the baseline wander, resulting in baseline 1. This corrected baseline 1 is then fed as input for a filter that calculates the median signal over every 50 subsequent data points. These median values are then subtracted from all the signals that are below 300 RFU (for ABI-devices) on baseline 1, resulting in baseline 2. This second baseline is often necessary due to the relatively short distance between the peaks that derive from probe products with only a few nucleotides difference. By applying this second baseline correction solely on the signals that are in the lower range of detection, even peaks that reside close to each other may reside back to zero-signal, without subtracting too much fluorescence that originates from the probe products. Program administrators can modulate the default baseline correction settings, and also may store different defaults for each used capillary system.

5.2.2 Peak detection

In capillary-based MLPA data analysis, peak detection is an essential step for subsequent analysis. Even though various peak detection algorithms for capillary electrophoresis data exist, most of them are designed for detection of peaks in sequencing profiles. While peak detection and peak size calling are very important processes for sequencing applications, peak quantification is not so important. Due to the relatively nature of the MLPA data, peak quantification is particularly important and has a large influence on the final results. Our peak detection algorithm exists of two separate steps; the first step exists of peak detection by comparison of the intensities of fluorescent units to set arbitrary thresholds and shape recognition, the second step exist of filtering of the generated peak list by relative comparison. Program administrators can modulate the peak detection algorithm thresholds, which make use of the following criteria:

1. Detection/Intensity threshold:
   This threshold is used to filter out small peaks in flat regions. The minimal and maximal peak amplitudes are arbitrary units and default values are provided for each different capillary system.
2. Peak area ratio percentage:
   Peak area is computed as the area under the curve within the distance of a peak candidate. Peak area ratio percentage is computed as the peak area divided by the total

amount of fluorescence times one hundred. The peak area ratio percentage of a peak must be larger than the minimum threshold and lower than the maximum set threshold.

3. Model-based criterion:
   The application of this criterion can consists of 3-4 steps:
   - Locate the start point for each peak: a candidate peak is recognized as soon as the signal increases above zero fluorescence.
   - Check if the candidate peak meets minimal requirements: the peak signal intensity is first expected to increase, if the top of the peak is reached and the candidate peak meets the set thresholds for peak intensity and peak area ratio percentage, then the peak is recognized as a true peak.
   - Discarding peak candidates: if the median signal of the previous 20 data points is smaller then the current peak intensity or if the current peak intensity returns to zero.
   - Detect the peak end: the signal is usually expected to drop back to zero designating the peak end. In some cases the signal does not return to zero, a peak end will therefore also be designated if the signal drops at least below half the intensity of the peak top and if the median signal of the 14 last data points is lower than the current signal.

4. Median signal peak filter:
   The median peak signal is calculated by the percentage of intensity of each peak as opposed to the median peak signal intensity of all detected peaks. Since the minimum and maximum thresholds are dependent on detected peaks, this filter will be applied after an initial peak detection procedure based on the criteria point 1-3.

5. Peak width filter:
   After peak end points have been identified, the peak width is computed as the difference of right end point and left end point. The peak width should be within a given range. This filter is also applied after an initial peak detection procedure.

6. Peak pattern recognition:
   This method is only applied for the size marker channel, and involves the calculation of the correlation between the data point of the peak top of the detected peak list (based on the criteria point 1-5) and the expected lengths of the set size marker. In case the correlation is less than 0.999, the previous thresholds will be automatically adapted and peak detected will be restarted. These adaptations mainly include adjustment of minimal and maximal threshold values.

5.2.3 Peak size calling

Size calling is a method that compares the detected peaks of a MLPA sample channels against a selected size standard. Lengths of unknown (probe) peaks can then be predicted using a regression curve between the data points and the expected fragment lengths of the used size standard, resulting in a fragment profile (figure 4). Coffalyser.NET allows the use of 2 different size-calling algorithms:

1. Local least squares method
2. 1st, 2nd or 3rd order least squares

The local least squares method is the default size calling method for our software. It determines the sizes of fragments (nucleotides) by using the local linear relationship between fragment length and mobility (data points). Local linearity is a property of functions that have graph that appear smooth, but they need not to be smooth in a mathematical sense. The local linear least squares method makes use of a function that is only once differentiable at a point where it is locally linear. Different from the other methods, this function is not differentiable, because the slope of the tangent line is undefined. To solve the local linear function our algorithm first calculates the intercept and coefficient for each size marker point of the curve by use of a moving predictor. A local linear size of 3 points provides three predictions for each point along its curve that is surrounded by at least 2 points. The average intercept and coefficient are then stored for that point. Points at the beginning and the end of the curve will receive a single prediction, since they do not have any surrounding known values. The coefficient (β) and intercept (α) are calculated by solving the following equations 1 and 2.

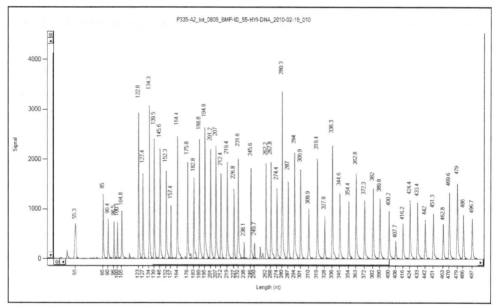

Fig. 4. MLPA fragment length profile displaying the lengths of all detected peaks from a sample. Peak lengths were determined by comparison of the data against a GS500-ABI size marker and determination of the length using the local least squares method.

$$\beta = \left(\frac{\sum X_i Y_i - \frac{1}{n}\sum X_i \sum Y_i}{\sum X_i^2 - \frac{1}{n}(\sum X_i)^2} \right) \qquad (1)$$

$$\alpha = \left(\overline{Y} - \left(\beta * \overline{X} \right) \right) \qquad (2)$$

E.g. if we use a size marker that has 15 known points and a local linear size of 3 points, the coefficient and intercept of point 5 will be calculated by equation 3 and 4.

$$\beta_5 = \frac{1}{3}\sum \frac{\sum X_{3-5}Y_{3-5} - \frac{1}{3}\sum X_{3-5}\sum Y_{3-5}}{\sum X_{3-5}^2 - \frac{1}{3}\left(\sum X_{3-5}\right)^2} ; \frac{\sum X_{4-6}Y_{4-6} - \frac{1}{3}\sum X_{4-6}\sum Y_{4-6}}{\sum X_{4-6}^2 - \frac{1}{3}\left(\sum X_{4-6}\right)^2} ; \frac{\sum X_{5-7}Y_{5-7} - \frac{1}{3}\sum X_{5-7}\sum Y_{5-7}}{\sum X_{5-7}^2 - \frac{1}{3}\left(\sum X_{5-7}\right)^2} \tag{3}$$

$$\alpha_5 = \frac{1}{3}\sum \left(\overline{Y} - \left(\beta_{3-5} * \overline{X}\right)\right); \left(\overline{Y} - \left(\beta_{4-6} * \overline{X}\right)\right); \left(\overline{Y} - \left(\beta_{5-7} * \overline{X}\right)\right) \tag{4}$$

To calculate the length of an unknown fragment our algorithm uses the calculated coefficient and intercepts calculated over the surrounded size marker peaks above and one below its peak. Each unknown point will be predicted twice where after the average value will be stored for that peak. If we wish to predict the length (Y) of an unknown fragment (X) of which the data point of the peak top is in between the data points of known fragments 5 and 6, we need to solve equation 5.

$$Y = \frac{1}{2}\sum \alpha_5 + \beta_5 * X; \alpha_6 + \beta_6 * X \tag{5}$$

5.2.4 Peak identification

Once all peaks have been size called, the profiles must be aligned to compare the fluorescence of the different targets across samples, an operation that is perhaps the single most difficult task in raw data analysis. Peaks corresponding to similar lengths of nucleotides may still be reported with slight differences or drifts due to secondary structures or bound dye compounds. These shifts in length make a direct numerical alignment based on the original probe lengths all but impossible. Our software uses an algorithm that automatically considers what the same peaks are between different samples, allowing easy peak to probe linkage. This procedure follows a window-based peak binning approach, whereby all peaks within a given window across different samples are considered to be the same peak (figure 5). Our software algorithm follows four steps: reference profile analysis, applying and prediction of new probe lengths, reiteration of profile analysis and data filtering of all samples.

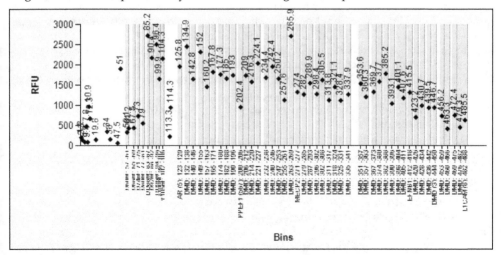

Fig. 5. Visualization of the collection of bins for a MLPA mix (x-axis) and the signal intensities in relative fluorescent units for detected peaks of a sample (y-axis).

The crucial task in data binning is to create a common probe length reference vector (or bin). In the first step our algorithm applies a bin set that searches for all peaks with a length closely resembling that of the design length of that probe. Next, the largest peak in each temporary bin is assumed to be the real peak descending from the related probe product. To create a stable bin, we calculate the average length over all real peaks of all used reference samples. If no reference samples exist, the median length over all collected real peak from all samples will be used. Since some probes may have a large difference between their original and detected length the previously created results may often not suffice. We therefore check if the length that we have related to each probe is applicable in our sample set. We do this by calculating how much variation exists over collected peaks length in each of the previous bins. If the variation was too large (standard deviation > 0.2) or no peak at all was found in any of the bins, the expected peak length for that probe will be estimated by prediction. The expected probe peak lengths may be predicted by using a second-order polynomial regression on using the available data of the probes for which reproducible data was found. Even though a full collection of bins is now available, the lengths of the probe products that were predicted may not be very accurate. The set of bins for each probe in the selected MLPA mix will therefore be improved by iteration of the previous steps. The lengths provided for the bins are now based on the previously detected or predicted probe product lengths allowing a more accurate detection of the real probe peaks. Probes that were not found are again predicted and a final length reference vector or bin is constructed for each probe. This final bin set can be used directly for data filtering but may also be edited manually in case the automatically created bin set may not suffice.

Data filtering is the actual process where the detected fragments of each sample are linked with gene information to a probe target or control fragment. Our algorithm assumes that peaks within each sample that fall within the same provided window or bin and have sufficient fluorescence intensity are the same probe (figure 4). Our algorithm is also able to link more than one peak to a probe within one sample. The amount of fluorescence of each probe product may then be expresses the peak height, peak area of the main peak and the summarized peak area of all peaks in a bin. An algorithm can then be used to compare these metrics and decide which should optimally be used as described at 3.2, alternatively users may set a default metric. The summarized peak area may reflect the amount of fluorescence best if peaks are observed that show multiple tops which all originate from the amplification of the same ligation product. Such peaks may be observed if:

1. Too much input DNA is added the amplification reaction and the polymerase was unable to complete the extension for all amplicons (Clark, J. M. 1988).
2. Peaks were discovered which are one base pair longer than the actual target due to non-template addition.
3. The polymerase was unable to complete the adenine addition on all products that resulted in the presence of shoulder peaks or +A/-A peaks (Applied Biosystems, 1988).

5.2.5 Raw data quality control

In the final step of the raw data analysis the software performs several quality checks and translates this into simple scores (figure 6).

These quality checks are the result of a comparison of sample specific properties such as: baseline height, peak signal intensity, signal to size drop, incorporated percentage of primer etc., to expected standards specific for each capillary system. Several quality checks are furthermore performed using the control fragments providing information about the used

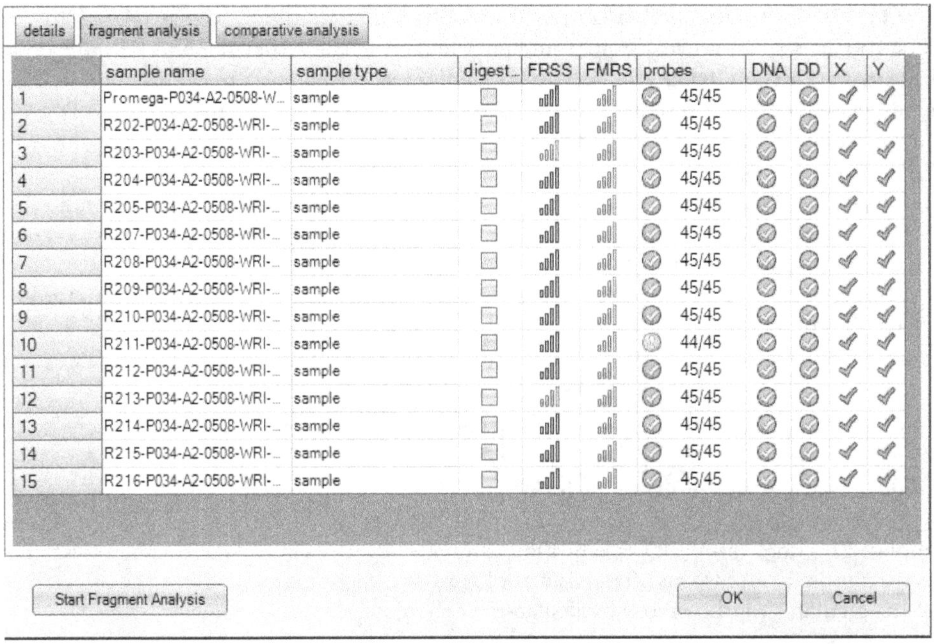

	sample name	sample type	digest	FRSS	FMRS	probes	DNA	DD	X	Y
1	Promega-P034-A2-0508-W...	sample	☐	⬛	⬛	⊘ 45/45	⊘	⊘	✓	✓
2	R202-P034-A2-0508-WRI-...	sample	☐	⬛	⬛	⊘ 45/45	⊘	⊘	✓	✓
3	R203-P034-A2-0508-WRI-...	sample	☐	⬛	⬛	⊘ 45/45	⊘	⊘	✓	✓
4	R204-P034-A2-0508-WRI-...	sample	☐	⬛	⬛	⊘ 45/45	⊘	⊘	✓	✓
5	R205-P034-A2-0508-WRI-...	sample	☐	⬛	⬛	⊘ 45/45	⊘	⊘	✓	✓
6	R207-P034-A2-0508-WRI-...	sample	☐	⬛	⬛	⊘ 45/45	⊘	⊘	✓	✓
7	R208-P034-A2-0508-WRI-...	sample	☐	⬛	⬛	⊘ 45/45	⊘	⊘	✓	✓
8	R209-P034-A2-0508-WRI-...	sample	☐	⬛	⬛	⊘ 45/45	⊘	⊘	✓	✓
9	R210-P034-A2-0508-WRI-...	sample	☐	⬛	⬛	⊘ 45/45	⊘	⊘	✓	✓
10	R211-P034-A2-0508-WRI-...	sample	☐	⬛	⬛	◐ 44/45	⊘	⊘	✓	✓
11	R212-P034-A2-0508-WRI-...	sample	☐	⬛	⬛	⊘ 45/45	⊘	⊘	✓	✓
12	R213-P034-A2-0508-WRI-...	sample	☐	⬛	⬛	⊘ 45/45	⊘	⊘	✓	✓
13	R214-P034-A2-0508-WRI-...	sample	☐	⬛	⬛	⊘ 45/45	⊘	⊘	✓	✓
14	R215-P034-A2-0508-WRI-...	sample	☐	⬛	⬛	⊘ 45/45	⊘	⊘	✓	✓
15	R216-P034-A2-0508-WRI-...	sample	☐	⬛	⬛	⊘ 45/45	⊘	⊘	✓	✓

Fig. 6. Coffalyser.NET screenshot. FRSS means fragment run separation score. FMRS means fragment MLPA reaction score. Probes, displays the number of found signals to the number of expected signals. The last columns display the quality of the DNA concentration and denaturation and the presence of the X and Y- fragments.

DNA itself as described before (Coffa, 2008). The quality scores allow users to easily find problems due to: the fragment separation process, MLPA reaction, DNA concentration or DNA denaturation. Users may then reject, accept and adjust sample types before starting the comparative analysis.

5.3 Comparative analysis

During the comparative part of the analysis we aim to isolate the amount of variation that was introduced over the repeated measured data and provide the user with meaningful data by means of reporting and visualization methods. The program is equipped with several normalization strategies in order to allow underlying characteristics of the different types of data sets to be compared. During normalization we bring MLPA data (probe peak signals) of unknown and reference samples to a common scale allowing easier understandable data to be generated. In MLPA, normalization refers to the division of multiple sets of data by a common variable or normalization constant in order to cancel out that variable's effect on the data. In MLPA kits, so called reference probes are usually added, which are targeted to chromosomal regions that are assumed to remain normal (diploid) in the DNA of all used samples.

Our algorithm is able to make use of the reference probes in multiple ways in order to comprise a common variable. In case a MLPA kit does not contain any reference probes, the common variable can be made out of probes selected by the user or the program will make an auto-selection. After normalization the relative amount fluorescence related to each probe can be expressed in dosage quotients, which is the usual method of interpreting MLPA data (Yau SC, 1996). This dosage quotient or ratio is a measure for the ratio in which the target sequence is present in the sample DNA as compared to the reference DNA, or relative ploidy. To make the normalization more robust our algorithm makes use of every MLPA probe signal, set as a reference probe for normalization to produce an independent ratio ($DQ_{i, h, j, z}$). The median of all produced ratios is then taken as the final probe ratio ($DQ_{i, h, j}$). This allows for the presence of aberrant reference signals without profoundly changing the outcome. If we want to calculate the dosage quotient for test Probe J of unknown Sample I as compared to t reference Sample H, by making use of reference Probes Z (1-n), we need to solve the equation 6.

$$DQ_{i,h,j} = med\left(\frac{\left[S_iP_j / S_iP_{z=1}\right]}{\left[S_hP_j / S_hP_{z=1}\right]}, \frac{\left[S_iP_j / S_iP_{z=2}\right]}{\left[S_hP_j / S_hP_{z=2}\right]}, \dots \frac{\left[S_iP_j / S_iP_{z=n}\right]}{\left[S_hP_j / S_hP_{z=n}\right]}\right) \qquad (6)$$

The data for each test probe of each sample ($DQ_{i, h, j}$) will be compared to each available reference sample ($S_h=n$), producing as many dosage quotients as there are reference samples. The final ratio ($DQ_{i, j}$) will then estimated by calculating the average over these dosage quotients. In case no reference samples are set, each sample will be used as reference and the median over the ratios be calculated.

5.3.1 Dealing with sample to sample variation

Each MLPA probe is multiplied during the amplification reaction with a probe specific efficiency that is mainly determined by the sequence of the probe, resulting in a probe specific bias. Even though the relative difference of these probes in signal intensity between different samples can be determined by normalization or visual assessment (figure 1), the calculated ratio results may not always be easy to understand by employing arbitrary thresholds only. This is mainly due to sample-to-sample variation or more specific, a difference in the amplification efficiency of probe targets between reference and sample targets. Chemical remnants from the DNA extraction procedure and other treatments sample tissue was subjected to, may allot to impurities that influence the *Taq* DNA polymerase fidelity. Alternatively target DNA sequences may have been modified by external factors, e.g. by aggressive chemical reactants and/or UV irradiation which may result in differences in amplification rate or extensive secondary structures of the template DNA that may prevent access to region of the target DNA by the polymerase enzyme (Elizatbeth van Pelt-Verkuil, 2008). An effect that is commonly seen with MLPA data is a drop of signal intensity that is proportional with the length of the MLPA product fragments (figure 7). This signal to size drop is caused by a decreasing efficiency of amplification of the larger MLPA probes and may be intensified by sample contaminants or evaporation during the hybridization reaction. Signal to size drop may further be influenced by injection bias of the capillary system and diffusion of the MLPA products within the capillaries.

In order to minimize the amount of variation in and between reference and sample data and create a robust normalization strategy our algorithm follows 7 steps. By automatic interpretation of results after each step our algorithm can adjust the parameters used for the next step thereby minimizing the amount of error that may be introduced by the use of

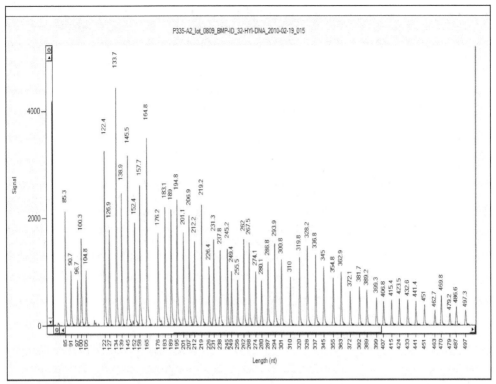

Fig. 7. MLPA fragment profile of a sample with a large drop in signal as size. This effect may have a similar result on the dosage quotients if not corrected for.

aberrant reference signals. The following 7 steps are performed in a single comparative analysis round:

1. Normalization of all data in population mode. Each sample will be applied as a reference sample and each probe will be applied as a reference probe.
2. Determination of significance of the found results by automatic evaluation using effect-size statistics and comparison of samples to the available sample type populations.
3. Measure of the relative amount of signal to size drop. If the relative drop is less than 12% a direct normalization will suffice, any larger drop will automatically be corrected by means of regression analysis (step 4-5).
4. Before correction of the actual amount of signal to size drop, samples are corrected for the MLPA mix specific probe signal bias. This can be done by calculating the extent of this bias in each reference run by regressing the probe signals and probe lengths using a local median least squares method. Correction factors for these probe specific biases are then computed by dividing the actual probe signal through its predicted signal. The final probe-wise correction factor is then determined by taking a median of the calculated values over all reference runs. This correction factor is then applied to all runs to reduce the effect of probe bias due to particular probe properties on the forthcoming regression normalization.

5. Next we calculate the amount of signal to size drop for every sample by using a function where the log-transformed probe bias corrected signals are regressed with the probe lengths using a 2nd order least squares method. Signals from aberrant targets are left out of this function, by applying an outlier detection method that makes use of the results found at step 2 as well as correlation measurements of the predicted line. The signal to size corrected values can then be obtained by calculating the distance of each log transformed pre-normalized signal to its predicted signal.

6. Normalization of signal to size corrected data in the user selected mode and determination of significance of the found results.

Our algorithm then measures the amount variation that could not be resolved in the final normalization to aid in results interpretation and automatic sample classification. To measure the imprecision of the normalization constant, each time a sample is normalized against a reference, the median of absolute deviations (MAD $_{i, h, j}$) is calculated between the final probe ratio ($DQ_{i, h, j}$) and the independent dosage quotients using each reference probe ($DQ_{i, h, j, z}$). The average of all collected MAD $_{i, j}$ values over the samples are then average to estimate the final amount of variation introduced by the imprecision of reference probes. Our algorithm estimates the final MAD $_{i, j}$ for each probe J in sample I and by equation 7.

$$MAD_{i,j} = \frac{1}{N}\sum_{z=1}^{N} med^{m}{}_{z=1}\left(\left|DQ_{i,h,j,z} - DQ_{i,h,j}\right|\right) \tag{7}$$

Since the final probe ratio ($DQ_{i, j}$) for each probe in each sample is estimated by the average over the dosage quotients ($DQ_{i, h, j}$) that were calculated using each reference sample (equation 8), the amount variation that was introduced over the different samples is estimated by calculating the standard deviation over these probe ratios (equation 9).

$$\sigma_{i,j} = \frac{1}{N}\sum_{h=1}^{N}\left(DQ_{i,h,j} - DQ_{i,j}\right)^{2} \tag{8}$$

$$DQ_{i,j} = \frac{1}{N}\sum_{h=1}^{N}\operatorname*{Med}_{z=1}^{M}\left(\frac{\left[S_{i}P_{j}/S_{i}P_{z}\right]}{\left[S_{h}P_{j}/S_{h}P_{z}\right]}\right) \tag{9}$$

Our algorithm then estimates the 95% confidence range of each probe ratio ($DQ_{i, j}$) of each sample by following 3 steps:

1. Conversion of the MAD values to standard deviations by multiplying with 1.4826 Albert, J. (2007)

2. Calculation of a single standard deviation for each probe ratio by combining the calculated value of step 1 with the standard deviation calculated over the reference samples by equation 9. This can be done by first converting both standard variations to variations by converting the values to the power of two. Then we sum up the outcome of both and take the square root.

3. Defining the limits of the confidence range by adding and subtracting a number of standard deviations of the final probe ratio ($DQ_{i, j}$) from equation 8.

Discrepancies on estimated dosage quotient by the used reference probes and/or reference samples may lead to an increase of the width of this confidence range, indicating a poor normalization. Since 95% is commonly taken as a threshold indicating virtual certainty (ZAR, J.H., 1984), our algorithm on default uses 1.96 standard deviations (equation 10) to calculate the confidence ranges for probe ratios.

$$\overset{95\%}{DQ}_{i,j} = +/-1.96 * \left(\sqrt{\left(1.4826*\left(MAD_{i,j}\right)\right)^2 + \left(\sigma_{i,j}\right)^2} \right) \tag{10}$$

5.3.2 Interpretation of the calculated dosage quotients

The previous sections explained how probe ratio are calculated and how our algorithm estimates the amount of introduced variation. In this section, we reflect on what those results mean for empirical comparison of users. To make data interpretation easier our program allows the use advanced visualization methods but also contains an algorithm allowing automatic data interpretation. Our algorithm compares the ratio and standard deviation of a test probe from a single sample to the behavior of that probe within a sub-collection of samples. This allows the program for instance to recognize if a result from an unknown sample is significantly different from the results found in the reference sample population. Alternatively, it may find if a sample is equal to a sample population, for instance a group of positive control samples. To make an estimation of the behavior of a probe ratio within a sample population, we calculate the average value and standard deviation for each probe over samples with the same sample type. In order to calculate the confidence range of probe J in for instance the reference sample population, we need to solve equation 11. N in this case refers to all probe ratio results ($DQ_{i, j}$) from samples that were defined in the normalization setup with the sample type: reference sample (h).

$$\overset{95\%}{DQ}_{ref\ j} = \overline{DQ_{i,j}} +/-1.96 * \left(\frac{1}{N}\sum_{i=h}^{N}\left(DQ_{i,j} - \overline{DQ_{i,j}}\right)^2 \right) \tag{11}$$

Probe result of each sample are then classified in three categories, by comparison to the confidence ranges of available sample types. A probe result of a sample is either significantly different to a sample population, equal to a sample population or the result is ambiguous. To define if a probe result of an unknown sample is significantly different (>>*) to sample population, our algorithm employs 2 criteria:
1. The difference in the magnitude of the probe ratio, as compared to the average of that probe calculated over samples with the same sample type, needs to exceed a delta value of 0.3. In case an unknown sample is compared to the reference sample population, the average ratio for each probe is always approaches 1.
2. The confidence range of the probe of the unknown sample (equation 10) cannot overlap with the confidence range of that probe in a sample population (equation 11).

An unknown sample in classified to be equal (=) to the population of a certain sample type if:
1. The difference in the magnitude of the probe ratio, as compared to the average of that probe calculated over samples with the same sample type, is less than 0.3.
2. The probe ratio of the unknown sample falls within the confidence range of that probe in a sample population (equation 11).

Probe results that are ambiguous, consequently only meet one of the two criteria in order to characterize the result to be different or equal. Ambiguous probe results that do show a difference in the magnitude of the probe ratio, as compared to the average of that probe calculated over samples with the same sample type, but have overlapping 95% confidence ranges will be marked with an asterisk (*). In case the overlap of the confidence ranges is less than 50% the probe results will be marked with a smaller or greater than symbol plus asterisk (<* or >*). Ambiguous probe results that do not show a difference in the magnitude of the probe ratio, but do show a difference in confidence ranges may be displayed with a single and double smaller or greater than symbols, depending on the size of the difference.

COFFALYSER/\/\/

Coffalyser MLPA sample report: P335-A2 lot 0809 BMP-ID 55-HYI-DNA
Sample type: Sample | Project: p335 | Experiment: p335 bmp
Performed by: Administrator | Machine: ABI | Report date: 10-8-2011 | Run date: 2-8-2011

D [nt]	Gene-Exon	Chr.band	Hg18 loc.	Probe Nr.	Height	Area	T-Area	Pre	Ratio	IRatio	[RPS]	[RPQ]	Stdev	[REF]	[Sam]	[Pos]	Width	E [nt]	d[nt]
59	Q60nt	.	.	.	0	0	0	0.00	0.00	0.00	0.00	0.00	0.00	.	.	.	0	0	0.0
66	Q68nt	.	.	.	0	0	0	0.00	0.00	0.00	0.00	0.00	0.00	.	.	.	0	0	0.0
73	Q74nt	.	.	.	0	0	0	0.00	0.00	0.00	0.00	0.00	0.00	.	.	.	0	0	0.0
79	Q80nt	.	.	.	0	0	0	0.00	0.00	0.00	0.00	0.00	0.00	.	.	.	0	0	0.0
85	DD88nt	.	.	.	1285	12957	12957	0.00	0.00	1.63	0.00	0.00	0.00	.	.	.	50	85.2	-0.1
91	Q92nt	.	.	.	789	8014	8014	0.00	0.00	1.00	0.00	0.00	0.00	.	.	.	49	90.4	-0.1
96	DD96nt	.	.	.	746	7663	7663	0.00	0.00	0.95	0.00	0.00	0.00	.	.	.	41	96.7	0.1
100	X100nt	.	.	.	733	7344	7344	0.00	0.00	0.93	0.00	0.00	0.00	.	.	.	41	100.1	-0.1
105	Y105nt	.	.	.	937	9527	9527	0.00	0.00	1.19	0.00	0.00	0.00	.	.	.	50	104.6	-0.1

Control fragments

D [nt]	Gene-Exon	Chr.band	Hg18 loc.	Probe Nr.	Height	Area	T-Area	Pre	Ratio	IRatio	[RPS]	[RPQ]	Stdev	[REF]	[Sam]	[Pos]	Width	E [nt]	d[nt]
454	EBF1-16	05q33.3	05-158057828	12527-13577	872	11996	11996	0.83	0.94	0.94	0.02	0.06	0.05	=	=	=	50	451.1	0.1
436	EBF1-14	05q33.3	05-158071810	13868-15386	1088	14794	14794	0.88	0.99	0.99	0.01	0.06	0.05	=	=	=	50	432.8	0.1
364	EBF1-10	05q33.3	05-158137017	14059-15657	1672	21110	21110	1.00	1.03	1.02	0.01	0.06	0.04	=	=	=	50	362.9	0.0
228	EBF1-1	05q33.3	05-158459187	12509-14269	1404	14904	14904	1.10	0.99	1.00	0.01	0.06	0.04	=	=	=	43	226.3	-0.1

Region: 05q
Chr. band: 05q33.3 Size: 401359 nt : 401 kb
Region analysis: Average 0.99 | Median 1.00 | Stdev 0.03 | Min 0.94 | Max 1.02

269	IKZF1-1	07p12.2	07-050315024	13877-15918	1866	20281	20281	1.04	0.97	0.95	0.01	0.06	0.04	=	=	=	50	267.3	0.0
208	IKZF1-2	07p12.2	07-050329206	14056-15654	2219	23996	23996	1.10	0.96	0.97	0.00	0.07	0.05	=	=	=	50	206.7	0.0
177	IKZF1-3	07p12.2	07-050337794	12536-13586	1879	19141	19141	1.10	0.93	0.93	0.03	0.06	0.05	=	=	=	48	176.1	0.0
263	IKZF1-4	07p12.2	07-050411797	13873-15917	1896	21431	21431	1.06	0.98	0.97	0.00	0.06	0.04	=	=	=	50	261.7	-0.1
142	IKZF1-5	07p12.2	07-050417749	13872-15390	2400	24318	24318	1.18	0.96	0.96	0.02	0.06	0.05	=	=	=	50	138.8	-0.1
465	IKZF1-6	07p12.2	07-050422507	14061-15923	678	9996	9996	0.85	0.98	1.00	0.01	0.06	0.05	=	=	=	50	462.9	0.2
337	IKZF1-7	07p12.2	07-050426950	13869-15387	2217	26815	26815	0.93	0.94	0.95	0.02	0.06	0.05	=	=	=	50	336.6	-0.2
288	IKZF1-8	07p12.2	07-050436025	12549-15921	1537	18376	18376	1.10	1.05	1.04	0.03	0.06	0.06	=	=	=	50	288.5	-0.1

Region: 07p
Chr. band: 07p12.2 Size: 121001 nt : 121 kb
Region analysis: Average 0.97 | Median 0.96 | Stdev 0.03 | Min 0.93 | Max 1.04

256	CDKN2A-5	09p21.3	09-021957822	10334-10871	225	3289	3289	0.18	0.17	0.17	0.00	0.01	0.01	<<*	<<*	?	26	255.3	0.0
251	CDKN2A-2a	09p21.3	09-021964958	10333-15916	297	2734	2734	0.24	0.22	0.23	0.01	0.01	0.01	<<*	<<*	?	21	249.2	0.0
238	CDKN2B-2	09p21.3	09-021995813	10337-15914	336	3298	3298	0.20	0.18	0.18	0.00	0.01	0.01	<<*	<<*	?	33	237.7	0.1
373	PAX5-10	09p13.2	09-036830475	12521-13571	1130	14355	14355	0.94	0.99	1.02	0.01	0.06	0.05	=	=	=	50	372.1	0.0
275	PAX5-8	09p13.2	09-036872066	14057-15919	1378	15004	15004	0.93	0.88	0.86	0.02	0.05	0.04	<<	<<	=	50	273.9	0.0
202	PAX5-6	09p13.2	09-036956622	14055-15653	2158	23552	23552	1.16	1.01	1.00	0.03	0.06	0.05	=	=	=	50	201.1	0.1
282	PAX5-5	09p13.2	09-036992699	13870-15920	3261	40183	40183	3.08	2.92	2.93	0.05	0.18	0.14	>>*	>>*	>>*	50	280	0.1
483	PAX5-2	09p13.2	09-037010669	13876-15924	1448	21743	21743	2.27	2.67	2.69	0.07	0.17	0.14	>>*	>>*	>>*	49	479.9	0.6
154	PAX5-1	09p13.2	09-037024272	12501-13551	1719	17812	17812	1.16	0.97	0.97	0.03	0.06	0.05	=	=	=	50	152.3	-0.1

Region: 09p
Chr. band: 09p13.2 - 09p21.3 Size: 15006 kb : 15 Mb
Region analysis: Average 1.12 | Median 0.97 | Stdev 0.97 | Min 0.17 | Max 2.93

301	ETV6-1	12p13.2	12-011694211	14058-15656	1766	21049	21049	1.06	1.02	1.05	0.01	0.06	0.05	=	=	=	47	300.5	-0.1
382	ETV6-1	12p13.2	12-011694328	14060-15658	1367	17495	17495	0.94	0.99	1.01	0.01	0.06	0.04	=	=	=	50	381.6	0.0
391	ETV6-2	12p13.2	12-011796704	13875-15393	1171	15572	15572	0.92	0.98	1.00	0.01	0.06	0.04	=	=	=	50	389.3	0.1
246	ETV6-3	12p13.2	12-011883341	13874-15915	1803	19440	19440	1.18	1.08	1.05	0.01	0.07	0.05	=	=	=	49	245.1	0.0
474	ETV6-5	12p13.2	12-011913651	13871-15389	1291	18192	18192	0.90	1.05	1.08	0.03	0.06	0.06	=	=	=	50	469.9	0.2
196	ETV6-8	12p13.2	12-011935363	14054-15652	2557	27619	27619	1.12	0.97	0.98	0.02	0.06	0.05	=	=	=	50	194.6	-0.1

Region: 12p
Chr. band: 12p13.2 Size: 241152 nt : 241 kb
Region analysis: Average 1.03 | Median 1.03 | Stdev 0.04 | Min 0.98 | Max 1.08

409	BTG1 AREA-up	12q21.33	12-090903053	14117-15718	357	4900	4900	0.37	0.41	0.41	0.01	0.03	0.02	<<*	<<*	<<*	44	407	0.1
160	BTG1 AREA-up	12q21.33	12-090905485	14116-15717	1071	11192	11192	0.51	0.42	0.43	0.01	0.03	0.02	<<*	<<*	<<*	45	157.5	-0.2
328	BTG1-2	12q21.33	12-091061700	12553-13603	862	10513	10513	0.43	0.43	0.43	0.00	0.03	0.02	<<*	<<*	<<*	50	327.8	-0.3
232	BTG1-1	12q21.33	12-091063286	12542-15913	1946	22529	22529	1.13	1.01	0.98	0.02	0.06	0.05	=	=	=	50	231.3	0.1

Region: 12q
Chr. band: 12q21.33 Size: 160233 nt : 160 kb
Region analysis: Average 0.56 | Median 0.43 | Stdev 0.24 | Min 0.41 | Max 0.98

220	RB1-6	13q14.2	13-047821213	01782-01346	1900	20475	20475	0.86	0.76	0.76	0.01	0.05	0.04	<<	<<	>>	50	219	-0.1
310	RB1-14	13q14.2	13-047851432	01789-01353	975	12043	12043	0.82	0.80	0.82	0.01	0.05	0.04	<<	=	>>	48	309.8	-0.1
355	RB1-19	13q14.2	13-047928373	01792-15922	1115	13537	13537	0.71	0.73	0.73	0.02	0.05	0.04	<<	<<	=	45	354.6	0.0
418	RB1-24	13q14.2	13-047945205	01797-01360	684	9239	9239	0.64	0.70	0.70	0.01	0.04	0.03	<<	<<	>>	50	415.5	0.2
445	RB1-26	13q14.2	13-047949488	01799-01362	773	10665	10665	0.66	0.74	0.75	0.02	0.04	0.04	<<	=	>>	50	441.7	0.2

Region: 13q
Chr. band: 13q14.2 Size: 128275 nt : 128 kb
Region analysis: Average 0.75 | Median 0.75 | Stdev 0.04 | Min 0.70 | Max 0.82

184	SHOX-AREA-down	Xp22.33P	X-000755205	06293-06219	1577	16612	17063	1.12	0.96	0.96	0.02	0.06	0.05	=	=	=	38	183	-0.1
148	SHOX-AREA-down	Xp22.33P	X-000770580	05648-06218	2171	22014	22014	1.30	1.06	1.07	0.03	0.06	0.05	=	=	=	50	145.3	-0.1
136	CRLF2-4	Xp22.33P	X-001281274	13889-15427	3065	31230	31230	1.34	1.08	1.09	0.03	0.07	0.05	=	=	=	50	133.6	-0.1
166	CSF2RA-16	Xp22.33P	X-001388386	14052-15650	2381	24386	24386	1.21	1.01	0.99	0.03	0.06	0.05	=	=	=	49	164.7	-0.2
346	IL3RA-1	Xp22.33P	X-001415701	13907-15445	1279	15854	15854	0.96	0.98	0.98	0.01	0.06	0.05	=	=	=	50	344.9	-0.1

Region: Xp
Chr. band: Xp22.33PAR Size: 660498 nt : 660 kb
Region analysis: Average 1.01 | Median 0.99 | Stdev 0.05 | Min 0.96 | Max 1.09

Fig. 8. Part of a pdf report from a tumor sample analyzed with the P335 MLPA kit. The report shows clear aberrations at 9p21.3, 9p13.2 and 12q21.33. Less clear is the ratio of RB1, which displays a slight decrease in signal as opposed to the reference population, but doesn't surpass the threshold value, due to sample mosaicism.

5.3.3 Reporting and visualization

Automatic data interpretation cannot replace the specialist judgment of a researcher. Knowledge about the expected genetic defect of the target DNA and other sample information may be crucial. To assist the user with data interpretation, our software automatically sorts all probe results based on the last updated map view locations of the probes. Chromosomal aberrations often-span larger regions (M. Hermsen, 2002), which allow probes targeted to that region to cluster together by sorting. Our software can then make a single page PDF reports, containing a summary of all relevant data, probe ratios (figure 8), statistics, quality controls and charts (figure 2 & 4) of a single sample.

	Gene	Positi...	Length	Order ▽	⊞ P335-A2_lo...	⊞ P335-A2_lo...	⊞ P335-A2_lo...	⊞ P335-A2_lo...
46	EBF1 - 16	05-158...	454	001	0.96	0.97	1.02	0.99
44	EBF1 - 14	05-158...	436	002	0.98	0.96	1	0.94
36	EBF1 - 10	05-158...	364	003	1.13	1.1	1.08	1.08
24	IKZF1 - 1	07-050...	269	004	1.08	1.08	0.55	1.11
14	IKZF1 - 2	07-050...	208	005	0.99	0.99	0.51	1.03
9	IKZF1 - 3	07-050...	177	006	0.94	0.8	0.48	0.86
23	IKZF1 - 4	07-050...	263	007	1.04	1.02	0.57	1.11
4	IKZF1 - 5	07-050...	142	008	1	0.95	0.49	0.9
47	IKZF1 - 6	07-050...	465	009	1.03	0.88	0.49	0.98
33	IKZF1 - 7	07-050...	337	010	0.91	0.92	0.56	0.92
27	IKZF1 - 8	07-050...	288	011	1.13	1.14	0.54	1.14
22	CDKN2A - 4	09-021...	256	012	0.61	0.88	0.99	0.98
21	CDKN2A - 2a	09-021...	251	013	0.64	1.06	0.94	1.01
19	CDKN2B - 2	09-021...	238	014	0.14	0.9	0.95	1.04
37	PAX5 - 10	09-036...	373	015	1.24	1.11	1	0.96
25	PAX5 - 8	09-036...	275	016	1.05	0.94	1.09	0.88
13	PAX5 - 6	09-036...	202	017	1.01	0.88	1.05	0.87
26	PAX5 - 5	09-036...	282	018	1.11	0.94	1.17	0.95
49	PAX5 - 2	09-037...	483	019	1.13	0.99	1.09	1.09
6	PAX5 - 1	09-037...	154	020	1.01	0.93	0.97	1.03
29	ETV6 - 1	12-011...	301	021	0.46	0.93	1.12	1
38	ETV6 - 1	12-011...	382	022	0.55	1.02	1.06	0.98
39	ETV6 - 2	12-011...	391	023	0.59	0.95	1.01	1.02
48	ETV6 - 5	12-011...	474	024	0.53	1.02	1.01	1.08
12	ETV6 - 8	12-011...	196	025	1.07	1.01	1.07	1.01
41	BTG1 AREA	12-090...	409	026	1.13	1.05	0.98	1
7	BTG1 AREA - 29	12-090...	160	027	0.99	1.01	1	0.96
32	BTG1 - 2	12-091...	328	028	1.08	1.01	1.04	1.11
18	BTG1 - 1	12-091...	232	029	1.08	0.96	1.04	0.96
16	RB1 - 6	13-047...	220	030	0.97	0.94	0.97	1
30	RB1 - 14	13-047...	310	031	0.97	0.95	0.97	0.97
35	RB1 - 19	13-047...	355	032	0.99	0.93	0.95	0.94
42	RB1 - 24	13-047...	418	033	1.01	0.94	1.02	1
45	RB1 - 26	13-047...	445	034	1	0.99	0.97	0.95
10	SHOX-AREA	23-755...	184	035	1.01	0.97	1.04	1.49
5	SHOX-AREA	23-770...	148	036	1.07	1.05	1.05	1.5
3	CRLF2 - 4	23-001...	136	037	1.06	0.99	1.07	1.46
8	CSF2RA - 16	23-001...	166	038	1.06	1.05	1.11	1.66
34	IL3RA - 1	23-001...	346	039	0.93	0.92	1.13	1.43
17	EBF1 - 1	05-158...	226	040	1.01	0.84	0.96	0.86
20	ETV6 - 3	12-011...	244	041	0.53	1.03	0.94	0.98
1	REFERENCE (125 nt)	02-242...	125	REF001	0.9	0.9	0.93	0.98
11	REFERENCE (190 nt)	03-038...	190	REF002	0.91	0.96	1.02	0.98
43	REFERENCE (427 nt)	05-013...	427	REF003	0.93	0.92	0.95	0.87
50	REFERENCE (490 nt)	05-090...	490	REF004	1.11	1.08	1.05	1.11
2	REFERENCE (130 nt)	05-132...	130	REF005	1.07	1.04	0.95	1.02
28	REFERENCE (294 nt)	07-072...	294	REF006	1.01	1.15	1.62	1.09
51	REFERENCE (500 nt)	11-001...	500	REF007	1	1	0.94	0.95
15	REFERENCE (214 nt)	11-020...	214	REF008	1.03	1.04	1.04	1.09
31	REFERENCE (319 nt)	15-046...	319	REF009	0.98	0.9	0.99	0.95
40	REFERENCE (400 nt)	20-025...	400	REF010	1.02	1.01	1.1	1.01

Fig. 9. Screen shot of from a tumor sample analyzed with the P335 MLPA kit. Probe ratio results of targets estimated as significantly increased as opposed to the reference population are marker green; those estimated as significantly decreased are marked red.

Our software enables users further, to display MLPA sample results in large array of different chart types figure 2 & 4). Charts may all be exported to different formats such as: jpg, gif, tiff, png, bmp. The results of a complete experiment may be plot together in grids and heat map algorithms may be applied to provide users a simple overview (figure 9). These grids may be exported to file formats (XML, txt, csv) that may be opened in Microsoft Excel. Alternatively these grids may also be exported to PDF files or several imaging formats.

6. Conclusions and future research

In this chapter we showed the options and applied algorithms of our MLPA analysis software, called Coffalyser.NET. Our software integrates new technologies enhancing the speed, accuracy and ease of MLPA analysis. Recognition of aberrations is improved by companioning effect-size statistics with statistical interference allowing users to interpreter units of measurement that are meaningful on a practical level (L. Wilkinson, 1999), while also being able to draw conclusions from data that are subject to random variation, for example, sampling variation (Bickel, Peter J.; Doksum, Kjell A., 2001). Our software contains extensive methods for results reporting and interpretation. It may also provide an alternative to software such as: Applied BioSystems Genotyper® and GeneScan® or GeneMapper®software; LiCor's SAGA, MegaBACE® Genetic Profiler and Fragment Profiler. Compatible with outputs from all major sequencing systems i.e. ABI Prism®, Beckman CEQ and MegaBACE® platforms. Coffalyser.NET is public freeware and can be downloaded from the MRC-Holland website.

Using data-mining techniques such as support vector machines in the large volumes of data obtained by large-scale MLPA experiments, may serve as a powerful and promising mechanism for recognizing of results patterns, which can be used for classification. Our future directions therefore concentrate on developing novel methods and algorithms that can improve recognition of disease related probe ratio patterns optimizing results in terms of validity, integrity and verification.

7. References

Ahn, J.W. (2007). Detection of subtelomere imbalance using MLPA: validation, development of an analysis protocol, and application in a diagnostic centre, BMC Medical Genetics, 8:9

Albert, J. (2007) Bayesian Computation with R. Springer, New York

Applied Biosystems. (1988). AmpFℓSTR® Profiler Plus™ PCR Amplification Kit user's manual.

Bickel, Peter J.; Doksum, et al. (2001). Mathematical statistics: Basic and selected topics. 1

Clark, J. M. (1988). Novel non-templated nucleotide addition reactions catalyzed by procaryotic and eucaryotic DNA polymerases. *Nucleic Acids Res* 16 (20): 9677–86.

Coffa, J. (2008). MLPAnalyzer: data analysis tool for reliable automated normalization of MLPA fragment data, Cellular oncology, 30(4): 323-35

Ellis, Paul D. (2010). The Essential Guide to Effect Sizes: An Introduction to Statistical Power, Meta-Analysis and the Interpretation of Research Results. United Kingdom: Cambridge University Press.

Elizatbeth van Pelt-Verkuil, Alex Van Belkum, John P. Hays (2008). Principles and technical aspects of PCR amplification.

González J. 2008. Probe-specific mixed model approach to detect copy number differences using multiplex ligation dependent probe amplification (MLPA), BMC bioinformatics, 9:261

Hermsen M., Postma C. (2002). Colorectal adenoma to carcinoma progression follows multiple pathways of chromosomal instability, Gastroenterology, 123 (1109-1119)

Holtzman NA, Murphy PD, Watson MS, Barr PA (1997). "Predictive genetic testing: from basic research to clinical practice". *Science (journal)* 278 (5338): 602–5.

Huang, C.H., Chang, Y.Y., Chen, C.H., Kuo, Y.S., Hwu, W.L., Gerdes, T. and Ko, T.M. (2007). Copy number analysis of survival motor neuron genes by multiplex ligation-dependent probe amplification. *Genet Med.* 4, 241-248.

Janssen, B., Hartmann, C., Scholz, V., Jauch, A. and Zschocke, J. (2005). MLPA analysis for the detection of deletions, duplications and complex rearrangements in the dystrophin gene: potential and pitfalls. *Neurogenetics.* 1, 29-35.

Kluwe, L., Nygren, A.O., Errami, A., Heinrich, B., Matthies, C., Tatagiba, M. and Mautner, V. (2005). Screening for large mutations of the NF2 gene. *Genes Chromosomes Cancer.* 42, 384-391.

Michils, G., Tejpar, S., Thoelen, R., van Cutsem, E., Vermeesch, J.R., Fryns, J.P., Legius, E. and Matthijs, G. (2005). Large deletions of the APC gene in 15% of mutation-negative patients with classical polyposis (FAP): a Belgian study. *Hum Mutat.* 2, 125-34.

Nakagawa, Shinichi; Cuthill, Innes C (2007). "Effect size, confidence interval and statistical significance: a practical guide for biologists". *Biological Reviews Cambridge Philosophical Society* 82 (4): 591–605

"NCBI: Genes and Disease". NIH: National Center for Biotechnology Information (2008).

Redeker, E.J., de Visser, A.S., Bergen, A.A. and Mannens, M.M. (2008). Multiplex ligation-dependent probe amplification (MLPA) enhances the molecular diagnosis of aniridia and related disorders. *Mol Vis.* 14, 836-840.

Schouten, J.P. (2002), Relative quantification of 40 nucleic acid sequences by multiplex ligation-dependent probe amplification. Nucleic Acids Research, 20 (12): e57

Scott, R.H., Douglas, J., Baskcomb, L., Nygren, A.O., Birch, J.M., Cole, T.R., Cormier-Daire, V., Eastwood, D.M., Garcia-Minaur, S., Lupunzina, P., Tatton-Brown, K., Bliek, J., Maher, E.R. and Rahman, N. (2008). Methylation-specific multiplex ligation-dependent probe amplification (MS-MLPA) robustly detects and distinguishes 11p15 abnormalities associated with overgrowth and growth retardation. *J Med Genet.* 45, 106-13.

Sequeiros, Jorge; Guimarães, Bárbara (2008). Definitions of Genetic Testing EuroGentest Network of Excellence Project.

Taylor, C.F., Charlton, R.S., Burn, J., Sheridan, E. and Taylor, GR. (2003). Genomic deletions in MSH2 or MLH1 are a frequent cause of hereditary non-polyposis colorectal cancer: identification of novel and recurrent deletions by MLPA. *Hum Mutat.* 6, 428-33.

Wilkinson, Leland; APA Task Force on Statistical Inference (1999). "Statistical methods in psychology journals: Guidelines and explanations". *American Psychologist* 54: 594–604. doi:10.1037/0003-066X.54.8.594.

Yau SC, Bobrow M, Mathew CG, Abbs SJ (1996). "Accurate diagnosis of carriers of deletions and duplications in Duchenne/Becker muscular dystrophy by fluorescent dosage analysis". *J. Med. Genet.* 33 (7): 550–558. doi:10.1136/jmg.33.7.550.

Zar, J.H. (1984) *Biostatistical Analysis.* Prentice Hall International, New Jersey. pp 43–45

6

Partitioning Error Sources for Quality Control and Comparability Analysis in Biological Monitoring and Assessment

James B. Stribling

Tetra Tech, Inc., Center for Ecological Sciences, Owings Mills, Maryland
USA

"...measurements are not passive accountings of an objective world but active interactions in which the thing measured and the way it is measured contribute inseparably to the outcome." (Lindley 2007: p. 154, attributing the concept to Neils Bohr)

"The experienced scientist has to learn to anticipate the possible sources of systematic error..." (Taylor 1997: p. 97)

"No simple theory tells us what to do about systematic errors. In fact, the only theory of systematic errors is that they must be identified and reduced...." (Taylor 1997: p. 106)

"...the only reason to carry out a test is to improve a process, to improve the quality..." (Deming 1986: p. i)

1. Introduction

Rationally, as scientists, we recognize that documented standard procedures constitute the first requirement for developing consistency within and among datasets; the second step is putting the procedures into practice. If the procedures were implemented as perfectly as they are written, there would be no need to question data. However, we are also cognizant of the fact that humans (a group of organisms to which we cannot deny holding membership) are called upon to use the procedures, and the consistency and rigor with which the procedures are applied are directly affected by an individual's skill, training, attention span, energy, and focus (Edwards, 2004). In fact, we fully expect inconsistency due to human foibles, and often substantial portions of careers are spent in efforts to recognize, isolate, correct, and minimize future occurrences of, error.

Many public and private organizations in the United States (US) and other countries collect aquatic biological data using a variety of sampling and analysis methods (Gurtz & Muir, 1994; ITFM, 1995a; Carter & Resh, 2001), often for meeting regulatory requirements, for example, by the United States' Clean Water Act (CWA) of 1972 (USGPO, 1989). While the information collected by an individual organization is usually directly applicable to a specific question or site-specific issue, the capacity for using it more broadly for comprehensive assessment has been problematic due to unknown data quality produced by different methods or programs (ITFM, 1995a; Diamond et al., 1996; NWQMC, 2001; The Heinz Center, 2002; GAO, 2004). If the occurrence and magnitude of error in datasets is unknown, a supportable conclusion based solely (or even in part) on those data is problematic at best. These datasets are more difficult to justify for analyses, communicate to broader audiences, base policy decisions on, and defend against potential misuse (Costanza et al., 1992; Edwards, 2004). To ensure the measurement system produces data that can be defended requires understanding the potential error sources that can affect variability of the data and approaches for monitoring the magnitude of error expression.

The purpose of this chapter is to communicate the concept of biological monitoring and assessment as a series of methods, each of which produces data and are as subject to error as any other measurement system. It will describe specific QC techniques and analyses that can be used to monitor variability (i.e., error), identify causes, and develop corrective actions to reduce or otherwise control error rates within acceptable limits. The chapter concludes by demonstrating that comparability analysis for biological data and assessment results is a two-step process, including 1) characterizing data quality, or the magnitude of error rates, associated with each method or dataset, and 2) determining acceptability. It should also be recognized that specific methods are not recommended in the chapter, but rather, emphasis is given that whatever methods are used, data quality and performance should be quantified. Additionally, special emphasis is given to biological monitoring where benthic macroinvertebrate sampling provides the primary data, but conceptually, this approach to QC is also applicable to other organism groups.

2. Quality control

Quality control (QC) is a process by which tests are designed and performed to document the existence and causes of error (=variability) in data, as well as helping determine what can be done to minimize or eliminate them, and developing, communicating, and monitoring corrective actions (CA). Further, it should also be possible to implement the QC process (Figure 1) in a routine manner such that, when those causes are not present, the cost of searching for them does not exceed budgetary constraints (Shewhart, 1939).

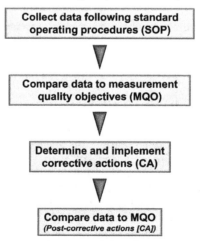

Fig. 1. Quality control (QC) process for determining the presence of and managing error rates, and thus, the acceptability of data quality.

The programmatic system that contains not only a series of QC tests and analyses, but also provides for organization and management of personnel, acquisition and maintenance of equipment and supplies essential to data collection, information management, information technology resources, safety protocols and facilities, enforcement of corrective actions, and budgetary support, is quality assurance (QA). It is acceptable to use the two terms jointly in reference to an overall quality program, as they often are, as QA/QC, but they should not be used interchangeably. The overall program is QA; the process for identifying and reducing error is QC.

Overall variability of data (= total uncertainty, or error) from any measurement system results from accumulation of error from multiple sources (Taylor 1988; Taylor & Kuyatt, 1994; Diamond et al., 1996; Taylor, 1997). Error can generally be divided into two types: systematic and random. Systematic error is the type of variability that results from a method and its application or mis-application; it is composed of bias that can, in part, be mediated by using an appropriate quality assurance program of training, audits, and documentation. Random error results from the sample itself or the population from which it is derived, and can only partly be controlled through a careful sampling design. It is often not possible to separate the effects of the two types of error, and they can directly influence each other (Taylor, 1988). The overall magnitude of error associated with a dataset is known as data quality; how statements of data quality are made and communicated are critical for data users and decision makers to properly evaluate the extent to which they should rely on technical, scientific, information (Keith, 1988; Peters, 1988; Costanza et al., 1992). Thus, an effective set of QC procedures helps not only reduce error in datasets, it provides tools for objective communication of uncertainty.

Biological assessment protocols are measurement systems consisting of a series of methods, each of which contribute to overall variability (Diamond et al., 1996; Cao et al., 2003; Brunialti et al., 2004; Flotemersch et al., 2006; Haase et al., 2006; Nichols et al., 2006; Blocksom & Flotemersch, 2008) (Figure 2). Our capacity as practitioners to control rates and magnitudes of error requires some attention be given to each component of the protocol. While it could be argued that error arising from any single component has only trivial effects on the overall indicator, lack of testing and documentation can substantially weaken that assertion, and opens the results to question. In fact, information without associated data quality characteristics might not even be considered data.

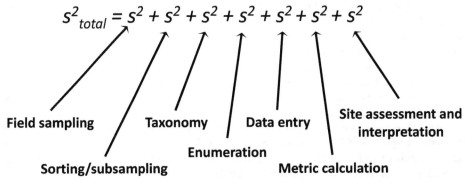

$$S^2_{total} = S^2 + S^2 + S^2 + S^2 + S^2 + S^2 + S^2$$

Field sampling   **Taxonomy**   **Data entry**   **Site assessment and interpretation**

Enumeration

Sorting/subsampling   **Metric calculation**

Fig. 2. Total error or variability (s^2) associated with a biological assessment is a combined result of that for each component of the process" (Flotemersch et al. 2006).

3. Indicators

All aquatic ecosystems are susceptible to cumulative impacts from human-induced disturbances including inorganic and organic chemical pollution, hydrologic alteration, channelization, overharvest, invasive species, and land cover conversion. Because they live in the presence of existing water chemistry and physical habitat conditions, the aquatic life of these systems (fish, insects, plants, shellfish, amphibians, reptiles, etc.) integrates cumulative effects of multiple stressors that are produced by both point and non-point source (NPS) pollution. The most common organism groups that are used by routine biological monitoring and assessment programs are benthic macroinvertebrates (aquatic insects, snails, mollusks, crustaceans, worms, and mites), fish, and/or algae, with indicators most often taking the form of a multimetric Index of Biological Integrity (IBI; Karr et al., 1986; Hughes et al., 1998; Barbour et al., 1999; Hill et al., 2000, 2003) or a predictive observed/expected (O/E) model based on the River Invertebrate Prediction and Classification System (RIVPACS; Clarke et al., 1996, 2003; Hawkins et al., 2000; Hawkins, 2006). Of these latter three groups, benthic macroinvertebrates (BM) are commonly used because the protocols are most well-established, the level of effort required for field sampling is reasonable (Barbour et al., 1999), and taxonomic expertise is relatively easily accessible. Thus, examples of QC tests and corrective actions discussed in this chapter are largely focused on benthic macroinvertebrates in the context of multimetric indexes, though, similar procedures for routine monitoring with algae and fish could be developed. Stribling et al. (2008) also used some of these procedures for documenting performance of O/E models.

4. Potential error sources in indicators

4.1 Field sampling

Whether the target assemblage is benthic macroinvertebrates, fish, or algae, the first step of biological assessment is to use standard field methods to gather a sample representing the taxonomic diversity and functional composition of a reach, zone, or other stratum of a waterbody. The actual dimensions of the sampling area ultimately depend on technical objectives and programmatic goals of the monitoring activity (Flotemersch et al., 2010). The spatial area from which the biological sample is drawn is that segment or portion of the waterbody the sample is intended to represent; for analyses and higher level interpretation, biological indicators are considered equivalent to the site. For its national surveys of lotic waters (streams and rivers), the U. S. Environmental Protection Agency defines a sample reach as 40x the mean wetted width (USEPA, 2004a); many individual states use a fixed 100m as the sampling reach.

Benthic macroinvertebrate samples are collected along 11 transects evenly distributed throughout the reach length, and a D-frame net with 500-μm mesh openings used to sample multiple habitats (Klemm et al., 1998; USEPA, 2004a; Flotemersch et al., 2006). An alternative approach to transects is to estimate the proportion of different habitat types in a defined reach (e.g., 100m), and distribute a fixed level of sampling effort in proportion to their frequency of occurrence throughout the reach (Barbour et al., 1999, 2006). For both approaches, organic and inorganic sample material (leaf litter, small woody twigs, silt, and sand) are composited in one or more containers, preserved with 95% denatured ethanol, and delivered to laboratories for processing. A composite sample over multiple habitats in a reach is a common protocol feature of many monitoring program throughout the US (Carter & Resh, 2001).

4.2 Laboratory processing

Processing of benthic macroinvertebrate samples is a 3-step process.   Sorting and subsampling serves to 1) isolate individual organisms from nontarget material, such as leaf litter and other detritus, bits of woody material, silt, and sand, and 2) prepare the sample (or subsample) for taxonomic identification. Taxonomic identification serves to match nomenclature to specimens in the sample, and enumeration provides the actual counts, by taxon, of everything contained within the sample.

Although it is widely recognized that subsampling helps to manage the level of effort associated with bioassessment laboratory work (Carter & Resh, 2001), the practice has been the subject of much debate (Courtemanch, 1996; Barbour & Gerritsen, 1996; Vinson & Hawkins, 1996). Fixed organism counts vary among monitoring programs (Carter & Resh, 2001), with 100, 200, 300 and 500 counts being most often used (Barbour et al., 1999; Cao & Hawkins, 2005; Flotemersch et al., 2006). Flotemersch & Blocksom (2005) concluded that a 500-organism count was most appropriate for large/nonwadeable river systems, based on examination of the relative increase in richness metric values (< 2%) between successive 100-organism counts. However, they also suggested that 300-organism count is sufficient for most study needs. Others have recommended higher fixed counts, including a minimum of 600 in wadeable streams (Cao & Hawkins, 2005). The subsample count used for the USEPA national surveys is 500 organisms (USEPA, 2004b); many states use 200 or 300 counts.

If organisms are missed during the sorting process, bias is introduced in the resulting data. Thus, the primary goal of sorting is to completely separate organisms from organic and inorganic material (e.g., detritus, sediment) in the sample. A secondary goal of sorting is to provide the taxonomist with a sample for which the majority of specimens are identifiable. Note that the procedure described here assumes that the sorter and the taxonomist are different personnel. Although it is not the decision of the sorter whether an organism is identifiable, straightforward rules can be applied that minimize specimen loss. For example, "counting rules" can be part of the standard operating procedures (SOP) for both the sorting/subsampling and taxonomic identification, such as specifying what not to count:

- Non-benthic organisms, such as free-swimming gyrinid adults (Coleoptera) or surface-dwelling veliids (Heteroptera)
- Empty mollusk shells (Mollusca: Bivalvia and Gastropoda)
- Non-headed worm fragments
- Damaged insects and crustaceans that lack at least a head and thorax
- Incidental collections, such as terrestrial insects or aquatic vertebrates (fish, frogs or tadpoles, snakes, or other)
- Non-macroinvertebrates, such as copepods, cladocera, and ostracods
- Exuviae (molted "skins")
- Larvae or pupae where internal tissue has broken down to point of floppiness

If a sorter is uncertain about whether an organism is countable, the specimen should be placed in the vial and not added to the rough count total.

The sorting/subsampling process is based on randomly selecting portions of the sample detritus spread over a gridded Caton screen (Caton, 1991; Barbour et al., 1999; see also Figures 6-4a, b of Flotemersch et al., 2006 [note that an individual grid square is 6 cm x 6 cm, or 36 cm^2, *not* 6 cm^2 as indicated in Figure 6-4b]). Prior to beginning the sorting/subsampling process, it is important that the sample be mixed thoroughly and distributed evenly across the sorting tray to reduce the effect of organism clumping that may have occurred in the sample container. The grids are randomly selected, individually removed from the screen, placed in a sorting tray, and all organisms removed with forceps;

the process is completed until the rough count by the sorter exceeds the target subsample size. There should be at least three containers produced per sample, all of which should be clearly labeled: 1) subsample to be given to taxonomist, 2) sort residue to be checked for missed specimens, and 3) unsorted sample remains to be used for additional sorting, if necessary.

The next step of the laboratory process is identifying the organisms within the subsample. A major question associated with taxonomy for biological assessments is the hierarchical target levels required of the taxonomist, including order, family, genus, species or the lowest practical taxonomic level (LPTL). While family level is used effectively in some monitoring programs (Carter & Resh 2001), the taxonomic level primarily used in most routine monitoring programs is genus. However, even with genus as the target, many programs often treat selected groups differently, such as midges (Chironomidae) and worms (Oligochaeta), due to the need for slide-mounting. Slide-mounting specimens in these two groups is usually (though, not always) necessary to attain genus level nomenclature, and sometimes even tribal level for midges. Because taxonomy is a major potential source of error in any kind of biological monitoring data sets (Stribling et al., 2003, 2008a; Milberg et al., 2008; Bortolus, 2008), it is critical to define taxonomic expectations and to treat all samples consistently, both by a single taxonomist and among multiple taxonomists. This, in part, requires specifying both hierarchical targets and counting rules. An example list of taxonomic target levels is shown in Table 1. These target levels define the level of effort that should be applied to each specimen. If it is not possible to attain these levels for certain specimens due to, for example, the presence of early instars, damage, or poor slide mounts, the taxonomist provides a more coarse-level identification. When a taxonomist receives samples for identification, depending upon the rigor of the sorting process (see above), the samples may contain specimens that either cannot be identified, or non-target taxa that should not be included in the sample. The final screen of sample integrity is the responsibility of the taxonomist, who determines which specimens should remain unrecorded (for any of the reasons stated above). Beyond this, the principal responsibility of the taxonomist is to record and report the taxa in the sample and the number of individuals of each taxon. Programs should use the most current and accepted keys and nomenclature. *An Introduction to the Aquatic Insects of North America* (Merritt et al., 2008) is useful for identifying the majority of aquatic insects in North America to genus level. By their very nature, most taxonomic keys are obsolete soon after publication; however, research taxonomists do not discontinue research once keys are available. Thus, it is often necessary to have access to and be familiar with ongoing research in different taxonomic groups. Other keys are also necessary for non-insect benthic macroinvertebrates that will be encountered, such as Oligochaeta, Mollusca, Acari, Crustacea, Platyhelminthes, and others. Klemm et al. (1990) and Merritt et al. (2008) provide an exhaustive list of taxonomic literature for all major groups of freshwater benthic macroinvertebrates. Although it is not current for all taxa, the integrated taxonomic information system (ITIS; http://www.itis.usda.gov/) has served as a clearinghouse for accepted nomenclature, including validity, authorship and spelling.

4.3 Data entry

Taxonomic nomenclature and counts are usually entered into the data management system directly from handwritten bench or field sheets. Depending on the system used, there may be an autocomplete function that helps prevent misspellings, but which can also contribute to errors. For example, entering the letters 'hydro' could potentially autocomplete as either

Hydropsyche or *Hydrophilus*, and the data entry technician on autopilot might continue as normal. There are also, increasingly, uses of e-tablets for entering field observation data, or direct entry of laboratory data into spreadsheets, obviating the need for hardcopy paper backup.

4.4 Data reduction/indicator calculation

There is a large number of potential metrics that monitoring programs can use (Barbour et al., 1999; Blocksom & Flotemersch, 2005; Flotemersch et al., 2006), requiring testing, calibration, and final selection before being appropriate for routine application. Blocksom & Flotemersch (2005) tested 42 metrics relative to different sampling methods, mesh sizes, and habitat types, some of which are based on taxonomic information, as well as stressor tolerance, functional feeding group, and habit. Other workers and programs have tested more and different ones. For example, the US state of Montana calibrated a biological indicator for wadeable streams of the "mountains" site class (Montana DEQ 2006), resulting in a multimetric index comprised of seven metrics (Table 2).

Taxon	Target
Ceratopogonidae	Ceratopogoninae, leave at subfamily; all others, genus level
Dolichopodidae	(Dolichopodidae)
Phoridae	(Phoridae)
Scathophagidae	(Scathophagidae)
Syrphidae	(Syrphidae)
Decapoda	Family
Hirudinea	Family
Hydrobiidae	(Hydrobiidae)
Nematoda	(Nematoda)
Nematomorpha	(Nematomorpha)
Nemertea	(Nemertea)
Turbellaria	(Turbellaria)
Chironomidae, the following genera are combined under *Cricotopus/Orthocladius*	
Cricotopus	*Cricotopus/Orthocladius*
Orthocladius	*Cricotopus/Orthocladius*
Cricotopus/Orthocladius	*Cricotopus/Orthocladius*
Orthocladius/Cricotopus	*Cricotopus/Orthocladius*
Chironomidae, the following genera are combined under *Thienemannimyia* genus group	
Conchapelopia	*Thienemannimyia* genus group
Rheopelopia	*Thienemannimyia* genus group
Helopelopia	*Thienemannimyia* genus group
Telopelopia	*Thienemannimyia* genus group
Meropelopia	*Thienemannimyia* genus group
Hayesomia	*Thienemannimyia* genus group
Thienemannimyia	*Thienemannimyia* genus group
Hydropsychidae, the following genera are combined under *Hydropsyche*	
Hydropsyche	*Hydropsyche*
Ceratopsyche	*Hydropsyche*
Hydropsyche/Ceratopsyche	*Hydropsyche*
Ceratopsyche/Hydropsyche	*Hydropsyche*

Table 1. In this example list of hierarchical target levels, all taxa are targeted for identification to genus level, unless otherwise noted. Taxa with target levels in parentheses are left at that level.

This discussion assumes that the indicator terms have already been calibrated and selected, and deals specifically with their calculation. For this purpose, the raw data are taxa lists and counts; their conversion into metrics is data reduction usually performed with computer spreadsheets or in relational databases.

To ensure that database queries are correct and result in the intended metric values, a subset of values should be recalculated by hand. One metric is calculated for all samples, all metrics are calculated for one sample. When recalculated values differ from those values in the matrix, the reasons for the disagreement are determined and corrections are made. Reports on performance include the total number of reduced values as a percentage of the total, how many errors were found in the queries, and the corrective actions specifically documented.

4.5 Indicator reporting

Regardless of whether the indicator is based on a multimetric framework or multivariate predictive model, the ultimate goal is to translate the quantitative, numeric result, the score, into some kind of narrative that provides the capacity for broad communication. The final assessment for a site is usually determined based on a site score relative to the distribution of reference site scores to reflect degrees of biological degradation, the more similar a test site is to reference less degradation is being exhibited. Depending on the calibration process and how many condition categories are structured, narratives for individual sites can come from two categories (degraded, nondegraded), three (good, fair, poor), four (good, fair, poor, very poor), or five (very good, good, fair, poor, or very poor). There also may be other frameworks a program chooses to use, but the key is to have the individual categories quantitatively-defined.

Metric	Description
Number of Ephemeroptera taxa	Count of the number of distinct taxa of mayflies in sample
Number of Plecoptera taxa	Count of the number of distinct taxa of stoneflies in sample
% individuals as EPT	Percent of individuals in sample that is mayflies, stoneflies, or caddisflies (Ephemeroptera, Plecoptera, or Trichoptera, respectively)
% individuals as non-insects	Percent of individuals in sample as non-insects
% individuals as predators	Percent of individuals in sample as predators
% of taxa as burrowers	Percent of taxa in sample as burrower habit
Hilsenhoff Biotic Index	Abundance-weighted mean of stressor tolerance values for taxa in the sample

Table 2. Sample-based metrics calculated for benthic macroinvertebrates. Shown are those developed and calibrated for streams in the "mountains" site class of the state of Montana, USA (Montana DEQ 2006, Stribling et al. 2008b).

5. Measurement quality objectives (MQO)

For each step of the biological assessment process there are different performance characteristics that can be documented, some of which are quantitative and others that are qualitative (Table 3). Measurement quality objectives (MQO) are control points above (or below) which most observed values fall (Diamond et al., 2006; Flotemersch et al., 2006; Stribling et al., 2003, 2008a, b; Herbst & Silldorf, 2006), and are roughly analogous to the Shewhart (1939) concept of process control.

Component method or activity	Performance characteristics				
	Precision	Accuracy	Bias	Representativeness	Completeness
1. Field sampling	●	na	Δ	Δ	●
2. Laboratory sorting/subsampling	●	na	●	Δ	●
3. Taxonomy	●	●	na	na	●
4. Enumeration	●	●	Δ	na	●
5. Data entry	na	●	na	na	●
6. Data reduction (e. g., metric calculation)	na	●	Δ	na	na
7. Site assessment and interpretation	●	●	Δ	Δ	●

Table 3. Error partitioning framework for biological assessments and biological assessment protocols for benthic macroinvertebrates. There may be additional activities and performance characteristics, and they may be quantitative (●), qualitative (Δ) or not applicable (na).

Specific MQO should be selected based on the distribution of values attained, particularly the minima and maxima. Importantly, for environmental monitoring programs, special studies should never be the basis upon which a particular MQO is selected; rather, they should reflect performance expectations when *routine* techniques and monitoring personnel are used. Consider MQO that are established using data from the best field team, or the taxonomist with the most years of experience, or the dissolved oxygen measurements taken using the most expensive field probes. When those people or equipment are no longer available to the program, how useful would the database be to future or secondary users? Defensibility would potentially be diminished. Values that are >MQO are not automatically taken to be unacceptable data points; rather, such values are targeted for closer scrutiny to determine possible reasons for exceedence and might indicate a need for corrective actions (Stribling et al. 2003, Montana DEQ 2006). Simultaneously, they can be used to help quantify performance of the field teams in consistently applying the methods.

5.1 Field sampling
Quantitative performance characteristics for field sampling are *precision* and *completeness* (Table 3). Repeat samples for purposes of calculating precision of field sampling are

obtained by sampling two adjacent reaches, shown as 500 m in this example (Figure 3), and can be done by the same field team for intra-team precision, or by different teams for inter-team precision. For benthic macroinvertebrates, samples from the adjacent reaches (also called duplicate or quality control [QC] samples) must be laboratory-processed prior to data being available for precision calculations. Assuming acceptable laboratory error, these precision values are statements of the consistency with which the sampling protocols 1) characterized the biology of the stream or river and 2) were applied by the field team, and thus, reflect a combination of natural variability and systematic error inherent in the dataset.

Primary reach (1°), 500m

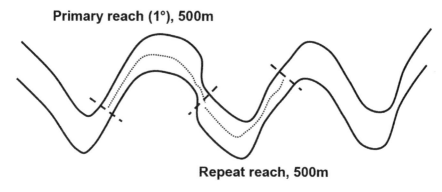

Repeat reach, 500m

Fig. 3. Adjacent reaches (primary and repeat) for calculating precision estimates (Flotemersch et al. 2006).

The number of reaches for which repeat samples are taken varies, but a rule-of-thumb is 10%, randomly-selected from the total number of sampling reaches constituting a sampling effort (whether yearly, programmatic routine, or individual project). Because they are the ultimate indicators to be used in address the question of ecological conditions, the metric and index values are used to calculate different precision estimates. Root-mean square error (RMSE) (formula 1), coefficient of variability (CV) (formula 2), and confidence intervals (formula 3) (Table 4) are calculated on multiple sample pairs, and are meaningful in that context. Documented values for field sampling precision (Table 5) demonstrate differences among individual metrics and the overall multimetric index (Montana MMI; mountain site class). Relative percent difference (RPD) (formula 4) (Table 4) can have meaning for individual sample pairs. For example, for the composite index, median relative percent difference (RPD) was 8.0 based on 40 sample pairs (Stribling et al., 2008b). MQO recommendations for that routine field sampling for that biological monitoring program were a CV of 10% and a median RPD of 15.0. Sets of sample pairs having with CV>10% would be subjected to additional scrutiny to determine what might be the cause of increased variability. Similarly, individual RPD values for sample pairs would be more specifically examined.

Percent completeness (formula 5) (Table 3, 4) is calculated to communicate the number of valid samples collected as a proportion of those that were originally planned. This value serves as one summary of data quality over the dataset and it demonstrates an aspect of confidence in the overall dataset.

Also called standard error of estimate, **root mean square error (RMSE)** is an estimate of the standard deviation of a population of observations and is calculated by:

$$RMSE = \sqrt{\frac{\sum\limits_{j=1}^{k}\sum\limits_{i=1}^{n_j}(y_{ij}-\bar{y}_j)^2}{\sum df_{1...k}}}$$

(1)

where y_{ij} is the ith individual observation in group j, j = 1...k (Zar 1999). Lower values indicate better consistency; and are used in calculation of the **coefficient of variability (CV)**, a unit-less measure, by the formula:

$$CV = \frac{RMSE}{\bar{Y}} \times 100$$

(2)

where $\bar{Y}$ is the mean of the dependent variable (e.g., metric, index across all sample pairs; Zar 1999). It is also known as relative standard deviation (RSD).

Confidence intervals (CI) (or detectable differences) are used to indicate the magnitude of separation of 2 values before the values can be considered different with statistical significance. A 90% significance level for the CI (i.e., the range around the observed value within which the true mean is likely to fall 90% of the time, or a 10% probability of type I error [α]). The 90% confidence interval (CI90) is calculated using RMSE by the formula:

$$CI90 = \pm([RMSE][z\alpha])$$

(3)

where za is the z-value for 90% confidence (i.e., p = 0.10) with degrees of freedom set at infinity. In this analysis, za = 1.64 (appendix 17 in Zar 1999). For CI95, the z-value would be 1.96. As the number of sample repeats increases, CI becomes narrower; we provide CI that would be associated with 1, 2, and 3 samples per site.

Relative percent difference (RPD) is the proportional difference between 2 measures, and is calculated as:

$$RPD = \left(\frac{|A-B|}{(A+B)/2}\right)x100$$

(4)

where A is the metric or index value of the 1st sample and B is the metric or index value of the 2nd sample (Keith, 1991; APHA, 2005; Smith, 2000). Lower RPD values indicate improved precision (as repeatability) over higher values.

Percent completeness (%C) is a measure of the number of valid samples that were obtained as a proportion of what was planned, and is calculated as:

$$\%C = \frac{v}{T}x100$$

(5)

where v is the number of valid samples, and T is the total number of planned samples (Flotemersch et al., 2006).

Percent sorting efficiency (PSE) describes how well a sample sorter has done in finding and removing all specimens from isolated sample material, and is calculated as:

$$PSE = \frac{A}{A+B} x100 \tag{6}$$

where A is the number of organisms found by the original sorter, and B is the number of missed organisms recovered (specimen recoveries) by the QC laboratory sort checker.

Percent taxonomic disagreement (PTD) quantifies the sample-based precision of taxonomic identifications by comparing target level taxonomic results from two independent taxonomists, using the formula:

$$PTD = \left[1 - \left(\frac{a}{N}\right)\right] \times 100 \tag{7}$$

where a is the number of agreements, and N is the total number of organisms in the larger of the two counts (Stribling et al., 2003, 2008a).

Percent difference in enumeration (PDE) quantifies the consistency of specimen counts in samples, and is determined by calculating a comparison of results from two independent laboratories or taxonomists using the formula:

$$PDE = \frac{|n_1 - n_2|}{n_1 + n_2} x100 \tag{8}$$

where n_1 is the number of organisms in a sample counted by the first laboratory, and n_2, the second (Stribling et al. 2003).

Percent taxonomic completeness (PTC) describes the proportion of specimens in a sample that meet the target identification level (Stribling et al. 2008) and is calculated as:

$$PTC = \frac{x}{N} x100 \tag{9}$$

where x is the number of individuals in a sample for which the identification meets the target level, and N is the total number of individuals in the sample.

Discrimination efficiency (DE) is an estimate of the accuracy of multimetric indexes and individual metrics is characterized as their capacity to correctly identify stressor conditions (physical, chemical, hydrologic, and land use/land cover) and is quantified as discrimination efficiency using the formula:

$$DE = \frac{a}{b} x100 \tag{10}$$

where a is the number of *a priori* stressor sites identified as being below the quantified biological impairment threshold of the reference distribution (25th percentile, 10th, or other), and b is the total number of stressor sites (Flotemersch et al., 2006).

Table 4. Explanations and formulas for quantifying 10 different performance characteristics for different steps of the biological assessment process.

Qualitative performance characteristics for field sampling are *bias* and *representativeness* (Table 3). Programs that use multihabitat sampling, either transect-based similar to that used by the US national surveys (USEPA 2004a), or distributing sampling effort among different habitat types (Barbour et al., 1999, 2006), are attempting to minimize the bias through two components of the field method. First, the approaches are not limited to one or a few habitat types; they are focused on sampling stable undercut banks, macrophyte beds, root wads, snags, gravel, sand, and/or cobble. Second, allocation of the sampling effort is distributed throughout the entire reach, thus preventing the entire sample from being taken in a shortened portion of the reach. Further, if the predominant habitat in a sample reach is poor or degraded, that habitat would be sampled as well. These field sampling methods are intended to depict the benthic macroinvertebrate assemblage that the physical habitat in the streams and rivers has the capacity to support. Another note about representativeness is to be cognizant that, while a method might effectively depict the property it is intended to depict (Flotemersch et al., 2006), it could be interpreted differently at different spatial scales (Figure 4).

Metric	RMSE	Mean	CV	CI90		
				1 sample	2 samples	3 samples
Number of Ephemeroptera taxa	0.94	5.25	17.9	1.55	1.1	0.89
Number of Plecoptera taxa	0.9	2.42	37.3	1.48	1.05	0.85
% individuals as EPT	8.86	47.98	18.5	14.53	10.27	8.39
% individuals as non-insects	3	7.3	41.1	4.93	3.49	2.85
% individuals as predators	5.32	16.91	31.4	8.72	6.17	5.03
% of taxa as burrowers	3.93	28.91	13.6	6.45	4.56	3.72
Hilsenhoff Biotic Index	0.47	4.27	10.9	0.76	0.54	0.44
Multimetric index (7-metric composite)	3.80	55.6	6.8	6.23	4.41	3.60

Table 5. Precision estimates for sample-based benthic macroinvertebrate metrics, and composite multimetric index (Stribling et al., 2008b). Data shown are from the US state of Montana, and performance calculations are based on 40 sample pairs from the "mountain" site class (*abbreviations* - RMSE, root mean square error; CV, coefficient of variation; CI90, 90 percent confidence interval; EPT, Ephemeroptera, Plecoptera, Trichoptera).

Fig. 4. Defining representativeness of a sample or datum first requires specifying the spatial and/or temporal scale of the feature it is intended to depict.

Accuracy is considered "not applicable" to field sampling (Table 3), because efforts to define analytical truth would necessitate a sampling effort excessive beyond any practicality. That is, the analytical truth would be all benthic macroinvertebrates that exist in the river (shore zone to 1-m depth). There is no sampling approach that will collect all individual benthic macroinvertebrate organisms.

5.2 Sorting/subsampling

Bias, precision, and, in part, *completeness,* are quantitative characteristics of performance for laboratory sorting and subsampling (Table 3). Bias is the most critical performance characteristic of the sorting process, and is evaluated by checking for specimens that may have been overlooked or otherwise missed by the primary sorter (Flotemersch et al., 2006). Checking of the sort residue is performed by an independent sort checker in a separate laboratory using the same procedures as primary, specifically, the same magnification and lighting as called for in the SOP. The number of specimens found by the checker as a proportion of the total number of originally found specimens is the percent sorting efficiency (PSE; formula 6) (Table 4), and quantifies sorting bias. This exercise is performed on a randomly-selected subset of sort residues (generally 10% of total sample lot), the selection of which is stratified by individual sorters, by projects, or by programs. As a rule-of-thumb, an MQO could be "less than 10% of all samples checked will have a PSE ≤90%". Table 6 shows PSE results from sort rechecks for a project within the state of Georgia (US). One sample (no. 8) exhibited a substantial failure with a PSE of 77.8, which became an immediate flag for a potential problem. Further evaluation of the results showed that the sample was fully sorted (100%), and still only 21 specimens were found by the original sorter, prior to the 6 recoveries by the re-check. Values for PSE become skewed when overall numbers are low, thus failure of this one sample did not indicate systematic error (bias) in the sorting process. Three additional samples fell slightly below the 90% MQO, but were only ≤ 0.2 percentage points low and were judged as passing by the QC analyst.

Precision of laboratory sorting is calculated by use of RPD with metrics and indexes as the input variables (Table 4). If, for example, the targeted subsample size is 200 organisms, and that size subsample is drawn twice from a sorting tray without re-mixing or re-spreading, metrics can be calculated from the two separate subsamples. RPD would be an indication of how well the sample was mixed and spread in the tray; the "serial subsampling" and RPD calculations should be done on two timeframes. First, these calculations should be done, and the results documented and reported to demonstrate what the laboratory (or individual sorter) is capable of in application of the subsampling method. Second, they should be done periodically to demonstrate that the program routinely continues to meet that level of precision. Representativeness of the sorting/subsampling process is addressed as part of the SOP that requires random selection of grid squares (Flotemersch et al., 2006) with complete sorting, until the target number is reached within the final grid. Percent completeness for subsampling is calculated as the proportion of samples with the target subsample size (±20%) in the rough sort. Considered as "not applicable", estimates of *accuracy* are not necessary for characterizing sorting performance.

5.3 Taxonomic precision (sample-based)

Precision and *completeness* are quantitative performance characteristics that are used for taxonomy (Table 3). Precision of taxonomic identifications is calculated using percent taxonomic

Sample no.	Number of specimens			PSE
	Original	Recovered	Total	
1	208	5	213	97.7
2	202	8	210	96.2
3	227	1	228	99.6
4	200	12	212	94.3
5	208	7	215	96.7
6	222	2	224	99.1
7	220	24	244	90.2
8	21	6	27	77.8[a]
9	215	22	237	90.7
10	220	25	245	89.8[b]
11	220	3	223	98.7
12	211	24	235	89.8[b]
13	205	12	217	94.5
14	213	24	237	89.9[b]
15	205	11	216	94.9
16	222	15	237	93.7
17	203	10	213	95.3
18	158	16	174	90.8

[a] Low PSE is due to there being small total number of specimens in the sample (n=27); this sample was also whole-pick (all 30 grid squares); [b] PSE values taken as passing, only ≤0.2 percentage points below MQO.

Table 6. Percent sorting efficiency (PSE) as laboratory sorting/ subsample quality control check. Results from 2006-2008 sampling for a routine monitoring program in north Georgia, USA.

disagreement (PTD) and percent difference in enumeration (PDE), both of which rely on the raw data (list of taxa and number of individuals) from whole-sample re-identifications (Stribling et al., 2003, 2008a). These two values are evaluated individually, and are used to indicate the overall quality of the taxonomic data. They can also be used to help identify the source of a problem. Percent taxonomic completeness (PTC) is calculated to document how consistently the taxonomist is able to attain the targeted taxonomic levels as specified in the SOP. It is important to note that the purpose of this evaluation approach is not to say that one taxonomist is correct over the other, but rather to make an effort to understand what is causing differences where they exist. The primary taxonomy is completed by one or more project taxonomists (T1); the re-identifications are completed as blind samples by one or more secondary, or QC taxonomists (T2) in a separate independent laboratory.

The number of samples for which this analysis is performed will vary, but 10% of the total sample lot (project, program, year, or other) is an acceptable rule-of-thumb. Exceptions are that large programs (>~500 samples) may not need to do >50 samples; small programs (<~30 samples) will likely still need to do at least 3 samples. In actuality, the number of re-identified samples will be program-specific and influenced by multiple factors, such as, how many taxonomists are doing the primary identification (there may be an interest in having 10% of the samples from each taxonomist re-identified), and how confident the ultimate data user is with the results. Mean values across all re-identified samples are estimates of taxonomic precision (consistency) for a dataset or a program.

5.3.1 Percent taxonomic disagreement (PTD)
The sample-based error rate for taxonomic identifications is quantified by calculation of percent taxonomic disagreement (PTD) (Table 4, formula 7). The key exercise performed by the QC analyst is determining the number of matches, or shared identifications between the two taxonomists (Table 7). Matches must be exact, that is, negative comparisons result even if the difference is *only* hierarchical (genus vs. family, or other), whether they have been assigned different names, or whether specimens are missing from the overall results of either T1 or T2. *Error typing* individual sample comparisons is the process of determining differences as either: a) straight disagreements, b) hierarchical differences, or c) missing specimens. While tedious, this QC exercise provides information that is extremely valuable in formulating corrective actions. An MQO of 15% has been found to be attainable by most programs, and is used for the USEPA national surveys. As testing continues and laboratories and taxonomists become more accustomed to the procedure, it is becoming apparent that potentially the national standard could eventually be set at 10%. A standard summary report for taxonomic identification QC (Table 8) can be effectively communicated to data users.

5.3.2 Percent difference in enumeration (PDE)
Another summary data quality indicator for performance in taxonomic identification is comparison of the total number of organisms counted and reported in the sample by the two taxonomists (not the sorters). There is some redundancy of this measure with PTD, but it has proven useful in helping highlight coarse differences immediately, and is calculated as percent difference in enumeration (PDE) (Table 4, formula 8). While sorters may be well-trained, experienced, and have substantial internal QC oversight, they may not always be able to determine identifiability, the final decision of which is the responsibility of the taxonomist. It is rare to find exact agreement on sample counts between two taxonomists but the differences are usually minimal, hence the low recommended MQO of 5%. When PDE>5, reasons are usually fairly obvious, and the QC analyst can turn attention directly to the error source to determine if it may be systematic, and the nature and necessity of corrective action(s).

5.3.3 Percent taxonomic completeness (PTC)
Percent taxonomic completeness (PTC) (Table 3, formula 9) quantifies the proportion of individuals in a sample that are identified to the specified target taxonomic level (Table 1). Results can be interpreted in a number of ways: the individuals in a sample are damaged or early instar, many are damaged with diagnostic characters missing (such as, gills, legs, antennae, etc.) or the taxonomist is inexperienced or unfamiliar with the particular taxon. MQO have not been used for this characteristic, but barring an excessively damaged sample, it is not uncommon to see PTC in excess of 97 or 98. For purposes of QC, it is more important to have the absolute difference (abs diff) of PTC between T1 and T2 to be a low number, as documentation of consistency of effort; those values are often typical at 5-6%, or below.

Sample no.	Count		No. matches	PDE	PTD	Target level (taxonomic completeness)				
	T1	T2				T1	PTC	T2	PTC	Abs diff
1	243	244	232	0.2	4.9	234	96.3	223	91.4	4.9
2	227	223	204	0.9	10.1	205	90.3	194	87.0	3.3
3	214	213	191	0.2	10.7	202	94.4	199	93.4	1.0
4	221	223	207	0.5	7.2	212	95.9	208	93.3	2.6
5	216	214	202	0.5	6.5	207	95.8	201	93.9	1.9
6	216	216	214	0	0.9	209	96.8	208	96.3	0.5
7	86	83	69	1.8	19.8	77	89.5	64	77.1	12.4
8	206	201	194	1.2	5.8	204	99	187	93.0	6.0
9	208	210	196	0.5	6.7	203	97.6	195	92.9	4.7
10	192	195	180	0.8	7.7	182	94.8	172	88.2	6.6

Table 7. Summary table for sample by sample taxonomic comparison results, from routine biological monitoring in US state of Mississippi. T1 and T2 are the primary and QC taxonomists, respectively. "No. matches" is the number of individual specimens counted and given the same identity by each taxonomist, and PDE, PTD, and PTC are explained in text. Target level is the number and percentage of specimens identified to the SOP-specified level of effort (see Table 3 as an example); "Abs diff" is the absolute difference between the PTC of T1 and T2.

A. Number of samples in lot	97
B. Number of samples used for taxonomic comparison	10
C. Percent of sample lot	10.3%
D. Percent taxonomic disagreement (PTD)	
1. MQO	15
2. No. samples exceeding	1
3. Average	7.9
4. Standard deviation	4.9
E. Percent difference in enumeration (PDE)	
1. MQO	5
2. No. samples exceeding	0
3. Average	0.6
4. Standard deviation	0.6
F. Percent taxonomic completeness (PTC, absolute difference)	
1. MQO	none
2. Average	4.3
3. Standard deviation	3.5

Table 8. Taxonomic comparison results from a bioassessment project in the US state of Mississippi.

5.4 Taxonomic accuracy (taxon-based)

Accuracy and *bias* (the inverse of accuracy) are quantitative performance characteristics for taxonomy (Table 3). Accuracy requires specification of an analytical truth, and for taxonomy that is 1) the museum-based type specimen (holotype, or other form of type specimen), 2) specimen(s) verified by recognized expert(s) in that particular taxon or 3) unique morphological characteristics specified in dichotomous identification keys. Determination of accuracy is considered "not applicable" for production taxonomy (most often used in routine monitoring programs) because that kind of taxonomy is focused on characterizing the sample; taxonomic accuracy, by definition, would be focused on individual specimens. Bias in taxonomy can result from use of obsolete nomenclature and keys, imperfect understanding of morphological characteristics, inadequate optical equipment, or poor training. Neither of these performance characteristics is considered necessary for production taxonomy, in that they are largely covered by the estimates of precision and completeness. For example, although it is possible that two taxonomists would put an incorrect name on an organism, it is considered low probability that they would put the *same incorrect name* on that organism.

5.5 Data entry accuracy

Recognition and correction of data entry errors (even the one mentioned in Section 4.3) could come from one of two methods for assuring accuracy in data entry; both do not need to be done. One is the double entry of all data by two separate individuals, and then performing a direct match between databases. Where there are differences, it is determined which database is in error, and corrections are made. The second approach is to perform a 100% comparison of all data entered to handwritten data sheets. Comparisons should be performed by someone other than the primary data entry person. When errors are found, they are hand-edited for documentation, and corrections are made electronically. The rates of data entry errors are recorded and segregated by data type (e.g., fish, benthic macroinvertebrates, periphyton, header information, latitude and longitude, physical habitat, and water chemistry). Issues could potentially arise when entering data directly into field e-tablets or laboratory computers. Because there would be no paper backup, QC checks of data entry are not possible.

5.6 Site assessment and interpretation

Quantitative performance characteristics for site assessment and interpretation are *precision*, *accuracy*, and *completeness* (Table 3). Site assessment precision is based on the narrative assessments from the associated index scores (good, fair, poor) from reach duplicates and quantifies the percentage of duplicate samples that are receiving the same narrative assessments. These comparisons are done for a randomly-selected 10% of the total sample lot. Table 9 shows this direct comparison that, for this dataset, 79% of the replicates returned assessments of the same category (23 out of 29); 17% were 1 category different (5 of 29); and 3% were 2 categories different (1 of 29). Assessment accuracy is expressed using discrimination efficiency (DE) (formula 10; Table 4), a value developed during the index calibration process, which relies upon, first, specifying magnitudes of physical, chemical, and/or hydrologic stressors that are unacceptable, and identifying those sites exhibiting those excessive stressor characteristics. The set of sites exhibiting unacceptable stressor levels constitute the analytical truth. The proportion of samples for which the biological index correctly identifies sites as impaired is DE. This is a performance characteristic that is directly suitable for expressing how well an indicator does what it is designed to do, detect stressor conditions, but it is not suitable for routine QC analyses. Percent completeness (%C) is the proportion of sites (of the total planned) for which valid final assessments were obtained.

6. Maintenance of data quality

The purpose of QC is to identify assignable causes of variation (error) so that the quality of the outcomes in future processes can be made, on average, less variable (Shewhart, 1939). For reducing error rates, it is first and foremost critical to know of the existence of error, and second, to know its causes. Once the causes are known, corrective actions can be designed to reduce or eliminate them. The procedures described in this chapter for gathering information that allow performance and data quality characteristics to be documented need to become a routine part of biological monitoring programs. If they are used only when "conditions are right", as part of special studies, or when there are additional resources, they are not serving their purpose and could ultimately be counter-productive. The counter-productivity would arise when monitoring staff begin to view QC samples and analyses as activities that are less than routine, and something for which to strive to do their best, that is, only when they are being tested. This perspective leads programs to work to meet a number, such as 15%, rather than using the information to maintain or improve.

Site	Replicate 1		Replicate 2		Categorical difference
	Narrative	Assessment category	Narrative	Assessment category	
A	Poor	3	Poor	3	0
B	Poor	3	Poor	3	0
C	Good	1	Good	1	0
D	Poor	3	Very Poor	4	1
E	Fair	2	Fair	2	0
F	Poor	3	Fair	2	1
G	Poor	3	Poor	3	0
H	Very Poor	4	Very Poor	4	0
I	Very Poor	4	Very Poor	4	0
J	Poor	3	Poor	3	0
K	Poor	3	Poor	3	0
L	Very Poor	4	Very Poor	4	0
M	Very Poor	4	Very Poor	4	0
N	Poor	3	Fair	2	1
O	Poor	3	Poor	3	0
P	Poor	3	Poor	3	0
Q	Poor	3	Very Poor	4	1
R	Poor	3	Poor	3	0
S	Fair	2	Very Poor	4	2
T	Fair	2	Fair	2	0
U	Good	1	Good	1	0
V	Poor	3	Fair	2	1
W	Fair	2	Fair	2	0
X	Poor	3	Poor	3	0
Y	Poor	3	Poor	3	0
Z	Very Poor	4	Very Poor	4	0
AA	Poor	3	Poor	3	0
BB	Fair	2	Fair	2	0
CC	Poor	1	Poor	1	0

Table 9. Assessment results shown for sample pairs taken from 29 sites, each pair representing two adjacent reaches (back to back (see Fig. 4). Assessment categories are 1 – good, 2 – fair, 3 – poor, and 4 – very poor.

Performance characteristic	MQO
Field sampling precision (multimetric index)	CV < 10%, for a sampling event (field season, watershed, or other strata)
Field sampling precision (multimetric index)	CI90 ≤ 15 index points, on a 100-point scale
Field sampling precision (multimetric index)	RPD < 15
Field sampling completeness	Completeness > 98%
Sorting/subsampling accuracy	PSE≥90, for ≥ 90% of externally QC'd sort residues
Taxonomic precision	Median PTD ≤ 15% for overall sample lot; samples with PTD ≥ 15% examined for patterns of error
Taxonomic precision	Median PDE ≤ 5%; samples with PDE ≥ 5% should be further examined for patterns of error
Taxonomic completeness	Median PTC ≥ 90%; samples with PTC ≤ 90% should be examined and those taxa not meeting targets isolated; mAbs diff ≤ 5%

Table 10. Key measurement quality objectives (MQO) that could be used to track maintenance of data quality at acceptable levels.

Key to maintaining data quality of known and acceptable levels is establishing performance standards based on MQO. Qualitative standards, such as some of the representativeness and accuracy factors (Table 3), can be evaluated by comparing SOP and SOP application to the goals and objectives of the monitoring program. However, a clear statement of data quality expectations, such as that shown in Table 10, will help to ensure consistency of success in implementing the procedures. As a program becomes more proficient and consistent in meeting the standards, efforts could be undertaken to "tighten up" the standards. With this comes necessary budgetary considerations; better precision can always be attained, but often at elevated costs.

7. Comparability analysis and acceptable data quality

All discussion to this point has been directed toward documenting data quality associated with monitoring programs, hopefully with sufficient emphasis that there are no data that are right or wrong, but just that they are acceptable or not. If data are acceptable for a decision (for example, in the context of biological assessment and monitoring), a defensible statement on the ecological condition of a site or an ecological system can be made. If they are not acceptable to support that decision, likewise, the decision not to use the data should also be defensible. Routine documentation and reporting of data quality within a monitoring program provides a statement of intra-programmatic consistency, that is, sample to sample comparability even if collected from different temporal or spatial scales. If there is an interest in or need to combine datasets from different programs (Figure 5), it is imperative for routinely documented performance characteristics be available for each. Lack of them will preclude any determination of acceptability for decision making by data users, whether scientists, policy-makers, or the public.

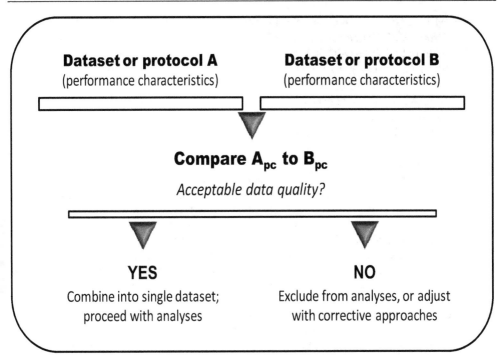

Fig. 5. Framework for analysis of comparability between or among monitoring datasets or protocols.

8. Conclusion

If data of unknown quality are used, whether by themselves or in combination with others, the assumption is implicit that they are acceptable, and hence, comparable. We must acknowledge the risk of incorrect decisions when using such data and be willing to communicate those risks to both data users and other decisionmakers. The primary message of this chapter is that appropriate and sufficient QC activities should be a routine component of any monitoring program, whether it is terrestrial or aquatic, focuses on physical, chemical, and/or biological indicators, and, if biological, whether it includes macroinvertebrates, algae/diatoms, fish, broad-leaf plants, or other organisms groups.

9. References

APHA. 2005. *Standard Methods for the Examination of Water and Wastewater.* 21st edition. American Public Health Association, American Water Works Association, and Water Environment Federation, Washington, DC.

Barbour, M.T., & J. Gerritsen. 1996. Subsampling of benthic samples: a defense of the fixed count method. *Journal of the North American Benthological Society* 15:386-391.

Barbour, M.T., J. Gerritsen, B.D. Snyder, J.B. Stribling. 1999. *Rapid Bioassessment Protocols for Streams and Wadeable Rivers: Periphyton, Benthic Macroinvertebrates and Fish.* Second edition. EPA/841-D-97-002. U.S. EPA, Office of Water, Washington, DC. *URL:* http://water.epa.gov/scitech/monitoring/rsl/bioassessment/index.cfm.

Barbour, M. T., J. B. Stribling, & P.F.M. Verdonschot. 2006. The multihabitat approach of USEPA's rapid bioassessment protocols: Benthic macroinvertebrates. *Limnetica* 25(3-4): 229-240.

Blocksom, K.A., & J.E. Flotemersch. 2005. Comparison of macroinvertebrate sampling methods for non-wadeable streams. *Environmental Monitoring and Assessment* 102:243-262.

Blocksom, K.A., & J.E. Flotemersch. 2008. Field and laboratory performance characteristics of a new protocol for sampling riverine macroinvertebrate assemblages. *River Research and Applications* 24: 373–387. DOI: 10.1002/rra.1073

Bortolus, A. 2008. Error cascades in the biological sciences: the unwanted consequences of using bad taxonomy in ecology. *Ambio* 37(2): 114-118.

Brunialti, G., P. Giordani, & M. Ferretti. 2004. Discriminating between the Good and the Bad: Quality Assurance Is Central in Biomonitoring Studies. Chapter 20, pp. 443-464, IN, G.B. Wiersma (editor), *Environmental Monitoring.* CRC Press.

Cao, Y., & C.P. Hawkins. 2005. Simulating biological impairment to evaluate the accuracy of ecological indicators. *Journal of Applied Ecology* 42:954-965.

Cao, Y., C.P. Hawkins, & M.R. Vinson. 2003. Measuring and controlling data quality in biological assemblage surveys with special reference to stream benthic macroinvertebrates. *Freshwater Biology* 48: 1898–1911.

Carter, J.L. & V.H. Resh. 2001. After site selection and before data analysis: sampling, sorting, and laboratory procedures used in stream benthic macroinvertebrate monitoring programs by USA state agencies. *Journal of the North American Benthological Society* 20: 658-676.

Caton, L. R. 1991. Improved subsampling methods for the EPA rapid bioassessment benthic protocols. *Bulletin of the North American Benthological Society* 8:317-319.

Clarke, R.T., M.T. Furse, J.F. Wright & D. Moss. 1996. Derivation of a biological quality index for river sites: comparison of the observed with the expected fauna. *Journal of Applied Statistics* 23:311–332.

Clarke, R.T., J.F. Wright & M.T. Furse. 2003. RIVPACS models for predicting the expected macroinvertebrate fauna and assessing the ecological quality of rivers. *Ecological Modeling* 160:219–233.

Costanza, R., S.O. Funtowicz & J.R. Ravetz. 1992. Assessing and communicating data quality in policy-relevant research. *Environmental Management* 16(1):121-131.

Courtemanch, D.L. 1996. Commentary on the subsampling procedures used for rapid bioassessments. *Journal of the North American Benthological Society* 15:381-385.

Deming, W.E. 1986. *Foreward.* In, Shewhart, W.A. 1939. *Statistical Methods from the Viewpoint of Quality Control.* The Graduate School, U.S. Department of Agriculture,

Washington, DC. 105 pp. Republished 1986, with a new Foreword by W.E. Deming. Dover Publications, Inc., 31 East 2nd Street, Mineola, NY.

Diamond, J.M., M.T. Barbour & J.B. Stribling. 1996. Characterizing and comparing bioassessment methods and their results: a perspective. *Journal of the North American Benthological Society* 15:713-727.

Edwards, P.N. 2004. "A vast-machine": Standards as social technology. *Science* 304 (7):827-828.

Flotemersch, J.E. & K.A. Blocksom. 2005 Electrofishing in boatable rivers: Does sampling design affect bioassessment *metrics? Environmental Monitoring and Assessment* 102:263-283. DOI: 10.1007/s10661-005-6026-2

Flotemersch J.E., J.B. Stribling, & M.J. Paul. 2006. *Concepts and Approaches for the Bioassessment of Non-Wadeable Streams and Rivers.* EPA/600/R-06/127. U.S. Environmental Protection Agency, Cincinnati, OH.

Flotemersch, J.E., J.B. Stribling, R.M. Hughes, L. Reynolds, M.J. Paul & C. Wolter. 2010. Site length for biological assessment of boatable rivers. *River Research and Applications.* Published online in Wiley InterScience (www.interscience.wiley.com) DOI: 10.1002/rra.1367.

General Accounting Office (GAO). 2004. *Watershed Management: Better Coordination of Data Collection Efforts.* GAO-04-382. Washington, DC , USA. Available from: <http://www.gao.gov/new.items/d04382.pdf>.

Gurtz, M.E. & T.A. Muir (editors). 1994. *Report of the Interagency Biological Methods Workshop.* U.S. Geological Survey, Open File Report 94-490, Reston, Virginia, USA.

Haase, P., J. Murray-Bligh, S. Lohse, S. Pauls, A. Sundermann, R. Gunn & R. Clarke. 2006. Assessing the impact of errors in sorting and identifying macroinvertebrate samples. *Hydrobiologia* 566:505–521. DOI 10.1007/s10750-006-0075-6

Hawkins, C.P. 2006. Quantifying biological integrity by taxonomic completeness: evaluation of a potential indicator for use in regional- and global-scale assessments. *Ecological Applications* 16:1277–1294.

Hawkins, C.P., R.H. Norris, J.N. Hogue & J.W. Feminella. 2000. Development and evaluation of predictive models for measuring the biological integrity of streams. *Ecological Applications* 10:1456–1477.

Heinz Center, The. 2002. *The state of the nation's ecosystems: measuring the lands, waters, and living resources of the United States.* The H. John Heinz III Center for Science, Economics, and the Environment, Washington, DC, USA. Cambridge University Press. Available from: <http://www.heinzctr.org/ecosystems/index.htm>.

Herbst, D.B. & E.L. Silldorf. 2006. Comparison of the performance of different bioassessment methods: similar evaluations of biotic integrity from separate programs and procedures. *Journal of the North American Benthological Society* 25:513–530.

Hill, B.H., A.T. Herlihy, P.R. Kaufmann, S.J. Decelles & M.A. Vander Borgh. 2003. Assessment of streams of the eastern United States using a periphyton index of biotic integrity. *Ecological Indicators* 2:325–338.

Hill, B.H., A.T. Herlihy, P.R. Kaufmann, R.J. Stevenson, F.H. McCormick & C.B. Johnson. 2000. Use of periphyton assemblage data as an index of biotic integrity. *Journal of the North American Benthological Society* 19:50–67.

Hughes, R.M., P.R. Kaufmann, A.T. Herlihy, T.M. Kincaid, L. Reynolds & D.P. Larsen. 1998. A process for developing and evaluating indices of fish assemblage integrity. *Canadian Journal of Fisheries and Aquatic Sciences* 55:1618–1631.

ITFM. 1995a. *The Strategy for Improving Water Quality Monitoring in the U.S.* Intergovernmental Task Force on Monitoring Water Quality. Report #OFR95-742, U.S. Geological Survey, Reston, Virginia, USA.

ITFM. 1995b. Performance-based approach to water quality monitoring. In: *Strategy for Improving Water Quality Monitoring in the U.S.*, Appendix M, Report #OFR95-742, Intergovernmental Task Force on Monitoring Water Quality, U.S. Geological Survey, Reston, Virginia, USA.

Karr, J.R., K.D. Fausch, P.L. Angermeier, P.R. Yant & I.J. Schlosser. 1986. *Assessing Biological Integrity in Running Waters: a Method and its Rationale.* Special publication 5. Illinois Natural History Survey, Champaign, Illinois, USA.

Keith, L.H. (editor). 1988. *Principles of Environmental Sampling.* ACS Professional Reference Book. American Chemical Society, Columbus, Ohio.

Keith, L.H. 1991. *Environmental Sampling and Analysis. A Practical Guide.* Lewis Publishers, Chelsea, Michigan.

Klemm, D.J., P.A. Lewis, F. Fulk & J.M. Lazorchak. 1990. *Macroinvertebrate Field and Laboratory Methods for Evaluating the Biological Integrity of Surface Waters.* EPA/600/4-90/030. Environmental Monitoring Systems Laboratory, U.S. Environmental Protection Agency, Cincinnati, OH. 256 pp.

Klemm, D.J., J.M. Lazorchak & P.A. Lewis. 1998. Benthic macroinvertebrates. Pages 147–182 in J. M. Lazorchak, D. J. Klemm, and D. V. Peck (editors). *Environmental Monitoring and Assessment Program – Surface Waters: Field Operations and Methods for Measuring the Ecological Condition of Wadeable Streams.* EPA/620/R-94/ 004F. U.S. Environmental Protection Agency, Washington, DC.

Lindley, D. 2007. *Uncertainty. Einstein, Heisenberg, Bohr, and the Struggle for the Soul of Science.* Anchor Books, a Division of Random House. ISBN: 978-1-4000-7996-4. New York, NY. 257 pp.

Merritt, R.W., K.W. Cummins & M.B. Berg (editors). 2008. *An Introduction to the Aquatic Insects of North America.* Fourth Edition. Kendall/Hunt Publishing Company, Dubuque, Iowa. ISBN 978-0-7575-5049-2. 1158 pp.

Milberg, P., J. Bergstedt, J. Fridman, G. Odell & L. Westerberg. 2008. Systematic and random variation in vegetation monitoring data. *Journal of Vegetation Science* 19: 633-644. http://dx.doi.org/10.3170/2008-8-18423.

Montana DEQ. 2006. *Sample collection, sorting, and taxonomic identification of benthic macroinvertebrates. Standard operation procedure. WQPBWQM-009. Revision no. 2.* Water Quality Planning Bureau, Montana Department of Environmental Quality, Helena, Montana. (Available from: http://www.deq.mt.gov/wqinfo/QAProgram/ WQPBWQM-009rev2_final_web.pdf)

NWQMC. 2001. *Towards a Definition of Performance-Based Laboratory Methods*. National Water Quality Monitoring Council Technical Report 01 – 02, U.S. Geological Survey, Reston, Virginia, USA.

Nichols, S.J., W.A. Robinson & R.H. Norris. 2006. Sample variability influences on the precision of predictive bioassessment. *Hydrobiologia* 572: 215–233. doi 10.1007/s10750-005-9003-4

Peters, J.A. 1988. Quality control infusion into stationary source sampling. Pages 317–333 in L. H. Keith (editor). *Principles of Environmental Sampling*. ACS Professional Reference Book. American Chemical Society, Columbus, Ohio.

Shewhart, W.A. 1939. *Statistical Methods from the Viewpoint of Quality Control*. The Graduate School, U.S. Department of Agriculture, Washington, DC. 105 pp. Republished 1986, with a new Foreword by W.E. Deming. Dover Publications, Inc., 31 East 2nd Street, Mineola, NY.

Smith, R.-K. 2000. *Interpretation of Organic Data*. ISBN 1-890911-19-4. Genium Publishing Corporation. Genium Group, Inc., Amsterdam, New York.

Stribling, J. B., S.R. Moulton II & G.T. Lester. 2003. Determining the quality of taxonomic data. *Journal of the North American Benthological Society* 22(4): 621-631.

Stribling, J.B., K.L. Pavlik, S.M. Holdsworth & E.W. Leppo. 2008a. Data quality, performance, and uncertainty in taxonomic identification for biological assessments. *Journal of the North American Benthological Society* 27(4): 906-919. doi: 10.1899/07-175.1

Stribling, J.B., B.K. Jessup & D.L. Feldman. 2008b. Precision of benthic macroinvertebrate indicators of stream condition in Montana. *Journal of the North American Benthological Society* 27(1):58-67. doi: 10.1899/07-037R.1

Taylor, J.K. 1988. *Defining the Accuracy, Precision, and Confidence Limits of Sample Data*. Chapter 6, pages 102-107, IN Lawrence H. Keith (editor), *Principles of Environmental Sampling*. ACS Professional Reference Book. ISBN 0-8412-1173-6. American Chemical Society. Columbus, Ohio.

Taylor, J.R. 1997. *An Introduction to Error Analysis. The Study of Uncertainties in Physical Measurements*. Second edition. University Science Books, Sausalito, California, USA.

Taylor, B.N. & C.E. Kuyatt. 1994. *Guidelines for Evaluating and Expressing the Uncertainty of NIST Measurement Results*. NIST Technical Note 1297. National Institute of Standards and Technology, U.S. Department of Commerce, Washington, DC. 24 pp.

USEPA. 2004a. *Wadeable Stream Assessment: Field Operations Manual*. EPA 841-B-04-004. Office of Water and Office of Research and Development, US Environmental Protection Agency, Washington, DC.

USEPA. 2004b. *Wadeable Stream Assessment: Benthic Laboratory Methods*. EPA 841-B-04007. Office of Water and Office of Research and Development, US Environmental Protection Agency, Washington, DC.

U. S. GPO (Government Printing Office). 1989. *Federal Water Pollution Control Act (33 U. S. C. 1251 et seq.)* as amended by a P. L. 92-500. In: Compilation of selected water resources and water pollution control laws. Printed for use of the Committee on Public Works and Transportation. Washington, DC, USA.

Vinson, M.R. & C.P. Hawkins. 1996. Effects of sampling area and subsampling procedure on comparisons of taxa richness among streams. *Journal of the North American Benthological Society* 15:392-399.

Zar, J.H. 1999. *Biostatistical Analysis. 4th edition.* Prentice Hall, Upper Saddle River, New Jersey, USA.

7

Application of Sampling Strategies for Hot-Mix Asphalt Infrastructure: Quality Control-Quality Assurance Sampling; Specification for Performance Test Requirements

Bor-Wen Tsai[1], Jiangmiao Yu[1,2] and Carl L. Monismith[1]
[1]University of California at Berkeley,
[2]South China University of Technology,
[1]U.S.A.
[2]P.R.C.

1. Introduction

Due to the lack of a rational, effective, and systematic quality control-quality assurance (QC/QA) methodology, the nonconformity of construction quality with design requirements for public works, especially for civil engineering infrastructure systems, can result in increased expenditures over time. Thus, development of a rational QC/QA methodology to ensure that the construction quality complies with the design requirements should have a high priority. The limited sample size constrained by the consideration of cost and time may result in the misjudgement that the construction quality does not meet the design requirements.

In this chapter, the effects of sampling size, sampling strategies, and acceptance/rejection criteria for QC/QA projects using statistically based decision making in hot-mix asphalt (HMA) construction are presented. Also, there has developed an increased interest recently in ensuring that the HMA as placed will meet certain performance requirements by measuring the actual performance parameters on test specimens prepared from in situ samples rather than from surrogate values such as asphalt content and aggregate gradation. Examples include direct measures of mix permanent deformation characteristics and fatigue characteristics, mix stiffness, and degree of compaction as measured by air-void content.

Determination of sample size is primarily based on an acceptable error level for a performance parameter specified by the agency. It is not uncommon to base quality assurance by many agencies on three samples. Through the *t* distributions, discussion is presented as to why it is not appropriate to take only this number of samples for quality-assurance. Based only on three samples in a large project, the agency will have insufficient power to reject the null hypothesis given that the null hypothesis is false unless the project quality delivered by the contractor is extremely poor so that the agency is confident enough to reject the project.

In addition to providing a general introduction to fundamental statistics and hypothesis testing, two case studies are used to clarify the relationships among sampling size, sample strategies, and performance specifications (or acceptance/rejection criterion). These include the following:

(1) A QC/QA case study is used to illustrate a methodology to determine strategies for a sampling scheme and selection of sample size for QC/QA for HMA construction to ensure that the acceptable level of a mix parameter is obtained with the same risk to the contractor and the agency. A sampling scheme and sampling size based on statistical simulation of a fixed length of a one-lane-width placement of HMA are discussed. Sample size is based on the combination of the sample size of the contractor and that of the agency to balance the risk to both organizations which will result in a mix that will meet the minimum performance requirement. An example is presented for the placement of 15,000 tons of HMA according to the California Department of Transportation (Caltrans) QC/QA requirements. For this total tonnage, the contractor and agency are assumed to perform a specific number of performance tests using the California stabilometer methodology for QC and QA.

(2) A QA case study is used to illustrate the application of the use of uniform design (UD) as a sampling strategy to ensure that the most representative sampling scheme can be achieved with a specified sample size. A sampling scheme using uniform design and sampling size through statistical simulation of a fixed length of a two-lane-width placement of HMA with several segregation data patterns is discussed. Based on the simulation, a QA guideline is developed by inspecting the accuracy of sample mean and the precision of sample standard deviation criteria combined with the application of the UD table is proposed and verified with two full scale pavement sections by measured air-void contents (measure of degree of compaction).

2. Case I: quality control-quality assurance sampling strategies for hot-mix asphalt construction

The effects of sampling strategies and size on statistically based decision making in hot-mix asphalt (HMA) construction are presented. For sample sizes agreed upon by the agency and the contractor, an acceptable level for an HMA mix parameter is determined with risk balanced between the two organizations. With increased emphasis on specific performance requirements, the use of performance tests on HMA specimens prepared from in situ samples is developing. Examples include direct measures of mix stiffness and permanent deformation characteristics. A measure of rutting resistance, the stabilometer S-value, is used by the California Department of Transportation (Caltrans) for quality control-quality assurance (QC/QA) projects. Although the S-value was used for this simulation because extensive tests were available, this approach is applicable to any performance measures already in use, such as HMA thickness or compacted air-void content. A sampling scheme and sampling size through statistical simulation of a fixed length of a one-lane-width placement of HMA are discussed. Sample size is based on the combination of the sample size of the contractor and that of the agency to balance the risk to both organizations and results in a mix that meets the minimum performance requirement.

2.1 Hypothesis testing of inequality

The acceptance or rejection of the null hypothesis, H_0, is referred to as a decision. Therefore, a correct decision is made in situations where (1) the H_0 is correctly accepted if H_0 is true and (2) the H_0 is correctly rejected if the H_0 is not true. As shown in the following for a decision based on a sample, when the null hypothesis is valid, the probability α of erroneously rejecting it is designated as the Type I error (or seller's risk), i.e., $\alpha = P\{\text{Type I error}\} =$

$P\{\text{reject } H_0 \mid H_0 \text{ is true}\}$; when the null hypothesis is not true, the probability β of erroneously accepting it is named the Type II error (or buyer's risk), i.e., $\beta = P\{\text{Type II error}\}$ = $P\{\text{fail to reject } H_0 \mid H_0 \text{ is false}\}$.

Truth about the population

	H_0 True	H_0 Not True
Reject H_0	Type I Error (α)	Correct Decision
Accept H_0	Correct Decision	Type II Error (β)

The power is defined as the probability $1 - \beta$ of correctly rejecting H_0 if H_0 is not true, i.e., $1 - \beta = P\{\text{reject } H_0 \mid H_0 \text{ is false}\}$. Hence, from the viewpoint of the agency (the buyer), it is necessary to have the power as high as possible; likely, from the perspective of the contractor (the seller), the Type I error should be as minimum as possible.

2.1.1 Testing inequality μ ≥ C$_s$ and size of test α

The objective is to test the null hypothesis that $H_0: \mu \geq C_S$ from the viewpoint of the contractor. The relevant t statistic is given by $t = \dfrac{\hat{\mu} - C_S}{SE(\hat{\mu})}$, where $\hat{\mu}$ is the sample mean of the stabilometer tests, C_S is the minimum specification limit for the stabilometer test, and SE the standard error. The critical region for the t test of size α of the null hypothesis is given by $t \leq -t_{1-\alpha,n-p}$, where $n = n_1 + \cdots + n_p$ and p the number of laboratories. In other words, the t statistic lies in the acceptance region $t > -t_{1-\alpha,n-p}$ if and only if

$$(\hat{\mu} - C_S)/SE(\hat{\mu}) \; > \; -t_{1-\alpha,n-p} \; \Rightarrow \; \hat{\mu} > C_S - t_{1-\alpha,n-p}SE(\hat{\mu}) \tag{1}$$

Note that the critical region for the t test of size $\alpha = 0.05$ of the null hypothesis $H_0: \mu \geq C_S$ can be given by $t \leq -t_{1-\alpha,n-p} \cong -\Phi(1-\alpha) = -\Phi(0.95) = -1.64485$, if $n-p \gg 0$, where Φ is the distribution function of a standard normal distribution. The size of test $\alpha = 0.05$ represents that at most a 5% chance is allowed to erroneously reject a valid null hypothesis; that is, there is a 95% chance that H_0 is accepted if H_0 is valid.

2.1.2 Test power, sample size, and operating-characteristic curve

Suppose that the hypothesis is not true, that is, $\mu < C_S$ (the opposite of $\mu \geq C_S$). Then the power as shown by Stone (Stone, 1996) is:

$$1 - \beta \cong P\left(\frac{\hat{\mu} - C_S}{SD(\hat{\mu})} \leq -z_{1-\alpha} \right)$$

$$= P\left(\frac{\hat{\mu} - \mu}{SD(\hat{\mu})} \leq -z_{1-\alpha} + \frac{C_S - \mu}{SD(\hat{\mu})} \right)$$

$$= \Phi\left(-z_{1-\alpha} + \delta \right)$$

where $\delta = \dfrac{|\mu - C_S|}{SD(\hat{\mu})}$ and $z_{1-\alpha}$ is quantile of a standard normal distribution. For the specified

α and β levels under the null hypothesis that $H_0 : \mu \geq C_S$, Table 1 lists the $SD(\hat{\mu})$, test power, and required sample sizes for the case with the agency and the contractor.

The test power equation shown in Table 1 indicates that the power of testing a null hypothesis is actually a standard normal distribution function in terms of the test of size α,

d ($d = \dfrac{|\mu - C_S|}{S_p}$), and the number of tests. Figure 1 plots power versus d with $\alpha = 0.05$ at

various numbers of tests, designated as the operating-characteristic curves. Several observations can be addressed in the following:

1. With the same number of tests and power level, increasing α will decrease the value of d; alternatively, at the same number of tests and a fixed value of d, increasing α will increase the power.
2. At fixed levels of α and β, increasing the number of tests will reduce the value of d.
3. For $n = 4$ to ensure that the test power is greater than 0.95, $d > 1.645$, i.e.,

    $d \cong \dfrac{|\hat{\mu} - C_S|}{S_S} > d_{0.95}$. In other words, if the sample mean of tests $\hat{\mu}$ is either

    $\hat{\mu} > C_S + d_{0.95} \cdot S_S$ or $\hat{\mu} < C_S - d_{0.95} \cdot S_S$, then the agency has enough power to confidently accept or reject the null hypothesis $H_0 : \mu \geq C_S$. If $\hat{\mu}$ lies in the range of $(C_S - d_{0.95} \cdot S_S, C_S + d_{0.95} \cdot S_S)$, then the agency does not have enough power with $n = 4$. Thus the number of tests has to be increased to reach the same level of power.
4. The test power approaches the test of size α as $d \to 0$.

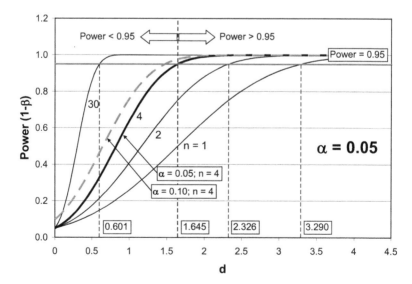

Fig. 1. Operating-characteristic curves with α = 0.05.

Application of Sampling Strategies for Hot-Mix Asphalt Infrastructure: Quality Control-Quality Assurance Sampling; Specification for Performance Test Requirements

129

The agency and the contractor:
$$n_1 = k \cdot n_2, \quad 0 < k \le 1$$
(n_1 : **number of QA samples;** n_2 : **number of QC samples.**)

$SD(\hat{\mu})$	$\dfrac{S_p}{p} \cdot \sqrt{\dfrac{1}{n_1} + \dfrac{1}{n_2}} = \dfrac{S_2}{2} \cdot \sqrt{\dfrac{n_1 + n_2}{n_1 \cdot n_2}}$
Test Power $1 - \beta$	$\Phi\left(-z_{1-\alpha} + 2 \cdot d \cdot \sqrt{\dfrac{k \cdot n_2}{1+k}}\right)$
Sample size	$n_2 \approx \dfrac{1+k}{k} \cdot \left(\dfrac{z_{1-\beta} + z_{1-\alpha}}{2 \cdot d}\right)^2$
Minimum Requirement of contractor	$\hat{\mu} \ge C_S - \dfrac{t_{1-\alpha, n_1 + n_2 - 2}}{2} \cdot \sqrt{\dfrac{1+k}{k \cdot n_2}} \cdot S_p$
Upper and Lower bounds of Agency	$\hat{\mu} \ge C_S + \dfrac{z_{1-\alpha} + z_{1-\beta}}{2} \cdot \sqrt{\dfrac{1+k}{k \cdot n_2}} \cdot S_p$ (upper bound)
	$\hat{\mu} \le C_S - \dfrac{z_{1-\alpha} + z_{1-\beta}}{2} \cdot \sqrt{\dfrac{1+k}{k \cdot n_2}} \cdot S_p$ (lower bound)

Note:

The pooled sample variance, S_p^2 , is defined as,

$$S_p^2 = \frac{\displaystyle\sum_{i=1}^{n_1}(x_{1,i} - \bar{x}_1)^2 + \sum_{i=1}^{n_2}(x_{2,i} - \bar{x}_2)^2 + \cdots + \sum_{i=1}^{n_p}(x_{p,i} - \bar{x}_p)^2}{n_1 + n_2 + \cdots + n_p - p} ; \text{ if } p = 1, \text{ then } S_p = S_S .$$

$z_{1-\alpha}$ and $z_{1-\beta}$ are quantiles of a standard normal distribution; Φ is the distribution function of a standard normal distribution.

$$d = \frac{|\mu - C_S|}{S_2} ; \quad \mu = \frac{\mu_1 + \mu_2}{2}$$

Table 1. Test power, required sample size, minimum requirement of contractor, and upper and lower bounds of agency.

2.1.3 Size of test α and power 1 - β

For the contractor, under the null hypothesis $H_0 : \mu \ge C_S$, the acceptance region for the t test of size α is given by Equation 1 , that is, $\hat{\mu} \ge C_S - t_{1-\alpha, n-p} \cdot SE(\hat{\mu})$.
For the agency, as noted earlier, the power of a test under the null hypothesis is given by

$$1 - \beta \cong \Phi(-z_{1-\alpha} + \delta) \implies z_{1-\beta} \cong -z_{1-\alpha} + \delta$$

where $\delta = \dfrac{|\mu - C_S|}{SD(\hat{\mu})} \cong \dfrac{|\hat{\mu} - C_S|}{SE(\hat{\mu})}$; $z_{1-\alpha}$ and $z_{1-\beta}$ are quantiles of a standard normal distribution.

Therefore, to satisfy the power requirement of the agency, $\delta \ge z_{1-\beta} - (-z_{1-\alpha})$, i.e.,

$$\frac{|\hat{\mu} - C_S|}{SE(\hat{\mu})} \geq z_{1-\alpha} + z_{1-\beta}$$

$$\Rightarrow \quad \hat{\mu} \geq C_S + (z_{1-\alpha} + z_{1-\beta}) \cdot SE(\hat{\mu}) \text{ , or } \hat{\mu} \leq C_S - (z_{1-\alpha} + z_{1-\beta}) \cdot SE(\hat{\mu}) \tag{2}$$

The $C_S + (z_{1-\alpha} + z_{1-\beta}) \cdot SE(\hat{\mu})$ will be designated as the upper bound and $C_S - (z_{1-\alpha} + z_{1-\beta}) \cdot SE(\hat{\mu})$ the lower bound of $1 - \beta$ power.

It should be noted that (1) if $\beta = 0.5$, then $z_{1-\beta} = 0$ and (2) $z_{1-\alpha} \cong t_{1-\alpha, n-p}$, if $n - p \gg 0$.

Thus, Equation 1 is equivalent to the lower bound of Equation 2. Based on Equation 1, the minimum requirement of the contractor, and Equation 2, the upper and lower bounds of power requirement of the agency, the case of the agency and the contractor is defined in Table 1.

Figure 2 illustrates plots of the upper and lower bounds at various power levels of the agency and the minimum requirement of the contractor under $H_0 : \mu \geq 37$ in terms of μ and sample size, n_2. The minimum requirements of the contractor in Figure 2 are plotted based on the t- distribution and standard normal distribution. It will be noted that the two curves coincide after $n_2 = 10$. From Table 1 and Figure 2, two observations can be made:

1. It is very important to recognize that the minimum requirement of the contractor actually matches the lower bound of 0.5 power of the agency.
2. The distance enclosed by the upper and lower bounds at a specified power level decreases with smaller S_P, larger α and β, larger k $(0 < k \leq 1)$, and, more importantly, larger sample size.

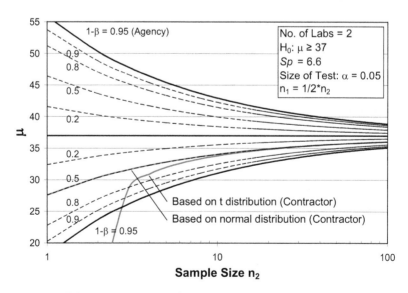

Fig. 2. Minimum stability requirements of the contractor and power requirement of the agency under the same null hypothesis.

2.2 QC/QA demonstration example

In this demonstration example 15,000 tons of HMA will be placed on 20 sublots (750 tons per sublot). The contractor is required to conduct 20 tests (n_2), i.e., one test per sublot. The number of tests conducted by the agency ($n_1 = k \cdot n_2$) will include the minimum required by the agency according to Caltrans specifications, i.e, $k = 0.1$ (2 tests in this case); in addition, determinations will be made for four tests ($k = 0.2$), six tests ($k = 0.3$), and eight tests ($k = 0.4$). The minimum stabilometer S-value has been set at 37 (Type A HMA) (California Department of Transportation [CALTRANS], 2007), and a standard deviation S_P is used for the S-value for tests between two laboratories of 6.6 (Paul Benson, private communication transmitting analyses of stabilometer test results for periods 1967– 1970 and 1995-1999). The demonstration example will include sampling consistency between QC and QA, sampling stabilization of S_P, and minimum requirements for both the agency and the contractor.

To conduct the sampling size simulation, several assumptions were made:
1. Lane width: 12 ft (3.66 m),
2. Unit weight of HMA – 145 lb/ft³ (2,323 kg/m³),
3. HMA layer thickness – 8 in. (20 cm), and
4. One stability sample is represented by a 4 × 4-in (10 × 10-cm) square with each square assigned a normalized stability value.

For these assumptions, the 15,000 tons of HMA will produce a section ~26,000 ft (7,925 m). long and 12 ft (3.66 m) wide. This results in a $N(0,1)$ stability population of 12 x 3 x 26,000 x 3 = 2,808,000 data points to generate three types of data patterns as schematically shown in Figure 3: (1) random pattern, (2) transverse strip pattern with 40 vertical strips, and (3) longitudinal strip pattern with 6 horizontal strips. The $N(0,1)$ distribution is separated by the points of quantiles into several intervals, e.g., 6 intervals for transverse strip pattern or 4 intervals for longitudinal strip pattern as shown in Figure 3. These intervals are then permuted to vary randomly across the x-direction or the y-direction of a lane of HMA paving. Those points within the interval are also randomly distributed over the transverse strip or the longitudinal strip.

The sampling scheme used was illustrated in Figure 4 with cases of $M \times N$ cells (N [y-direction] = 1; M [x-direction] = 10, 20, 30, 40, 50, 100, 200, and 500). That is, one random QC sample from each cell and one random QA sample from one random cell of $n_1 = k \cdot n_2$ random transverse strips. A total of 8 cases were simulated over three data patterns. Each case, per data pattern, was simulated 200 times.

To verify the minimum sampling size for an HMA paving strip is to show (1) no apparent difference of sampling consistency between the contractor (QC) and the agency (QA) and (2) stabilization of the pooled sample estimate of standard deviation of stability value, S_P. (Tsai & Monismith, 2009).

In each sampling simulation, the normalized stability values form a distribution with mean and standard deviation; hence, when repeated 200 times, the standard deviations will form another distribution. For each case, the standard deviations of the standard deviation distributions (SDSD) were calculated for QC and QA respectively. The difference of SDSD between QA and QC were used as an index to represent the sampling consistency between the agency (QA) and the contractor (QC).

Likewise, for each simulation, the S_p was calculated based on the equation in Table 1; hence, when repeated 200 times, the standard deviation of the S_p distribution will be used to inspect its stability over the $M \times N$ domain.

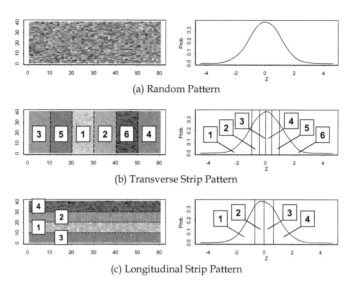

(a) Random Pattern

(b) Transverse Strip Pattern

(c) Longitudinal Strip Pattern

Fig. 3. Schemetic illustration of three data patterns: (a) random pattern, (b) transverse strip pattern, and (c) longitudinal strip pattern.

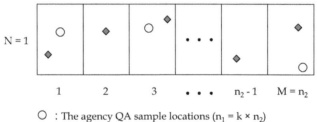

$\bigcirc$: The agency QA sample locations ($n_1 = k \times n_2$)

$\blacklozenge$: The contractor QC sample locations (n_2)

Fig. 4. Sampling scheme.

Figure 5a illustrates the simulation results for sampling consistency between QC and QA at various k values in terms of global smoothed line over three different data patterns. As would be expected, the sampling consistency between QC and QA increases as the k value increases. Figure 5b indicates that sampling stabilization of S_P depends only on the contractor's sampling size, n_2, rather than the k value.

From a series of operating-characteristic curves for the four k values and two α values (5% and 10%), the values in Table 2 were determined for the required minimum value of S, termed μ_{min}. With Figure 6a as an example, under the condition that $\alpha = 5\%$, $n_2 = 20$, $k = 0.2$, and power = 0.95, d has to be greater than 0.902 to satisfy the agency's power requirement; that is, $\hat{\mu}$ has to be greater than 42.95 so that the agency has power 0.95 to clearly accept the contractor's mix. Figure 6b shows a smaller d (0.803) will be obtained when the α value is increased to 10%.

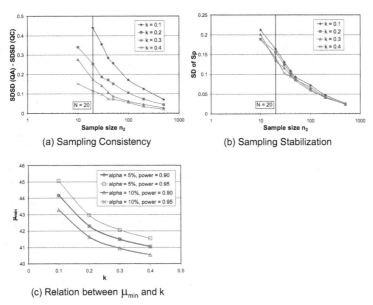

(a) Sampling Consistency   (b) Sampling Stabilization

(c) Relation between μ_{min} and k

Fig. 5. Summary of simulation results: (a) sampling consistency; (b) sampling stabilization; and (c) relationship between k and μ_{min}.

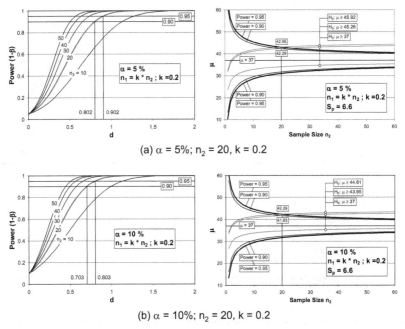

(a) α = 5%; n_2 = 20, k = 0.2

(b) α = 10%; n_2 = 20, k = 0.2

Fig. 6. Examples of operating-characteristic curves and μ_{min} required to meet the agency's power requirement and the contractor's minimum requirement: (a) α = 5% and (b) α = 10%.

Figure 5c illustrates the relationship between k and the minimum S-value. It is apparent that an increase of k-value reduces the value of μ_{min}. It is interesting to observe that the curve of $\alpha = 5\%$ and power = 0.90 is exactly the same as the curve of $\alpha = 10\%$ and power = 0.95. From Figure 5c, it is also shown that the higher μ_{min} -criterion is needed if both the agency and the contractor require a high power level and a low α-level, whereas if both the agency and the contractor require a low power level and a high α level, then the μ_{min} criterion can be much smaller.

α (Contractor)	Power 1-β (Agency)	k	d	Acceptance μ_{min}	Target Hypothesis (Contractor)
5%	0.90	0.1	1.085	44.16	$H_0^* : \mu \geq 48.19$
		0.2	0.802	42.29	$H_0^* : \mu \geq 45.26$
		0.3	0.680	41.49	$H_0^* : \mu \geq 44.02$
		0.4	0.613	41.05	$H_0^* : \mu \geq 43.31$
	0.95	0.1	1.220	45.05	$H_0^* : \mu \geq 49.07$
		0.2	0.902	42.95	$H_0^* : \mu \geq 45.92$
		0.3	0.766	42.06	$H_0^* : \mu \geq 44.58$
		0.4	0.688	41.54	$H_0^* : \mu \geq 43.81$
10%	0.90	0.1	0.951	43.27	$H_0^* : \mu \geq 46.41$
		0.2	0.703	41.63	$H_0^* : \mu \geq 43.95$
		0.3	0.597	40.94	$H_0^* : \mu \geq 42.91$
		0.4	0.537	40.54	$H_0^* : \mu \geq 42.31$
	0.95	0.1	1.086	44.16	$H_0^* : \mu \geq 47.30$
		0.2	0.803	42.29	$H_0^* : \mu \geq 44.61$
		0.3	0.682	41.50	$H_0^* : \mu \geq 43.46$
		0.4	0.613	41.05	$H_0^* : \mu \geq 42.81$

Note:

Null hypothesis: $H_0 : \mu \geq 37$.

$n_1 = k \cdot n_2 \ (0 \leq k \leq 1)$, where n_1 is the number of tests of agency; n_2 the number of tests of contractor.

$d = \dfrac{|\mu - C_S|}{S_P}$, where $\mu = \dfrac{\mu_1 + \mu_2}{2}$, μ_1 the average stabilometer value from agency; μ_2 the average stabilometer value from contractor; $C_S = 37$; $S_P = 6.6$.

$$\mu_{min} = d \cdot S_P + C_S.$$

Table 2. Acceptance μ_{min} values and target hypotheses for contractor with combinations of various α levels, power levels, and k- values.

3. Case II: HMA sampling strategies using uniform experimental design for quality assurance

The application of using uniform design (UD) as a sampling strategy for quality assurance (QA) ensures that the most representative and unbiased sampling scheme can be achieved with the sample size based on an acceptable error level of a hot-mix asphalt (HMA) parameter specified by the agency. Through statistical simulations and demonstration of air-void measurements of two field pavement sections, a QA guideline combined with the UD sampling scheme was developed to justify construction quality using the sample mean and sample standard deviation criteria. This approach can also be applied to any performance measure already in use.

3.1 Uniform experimental design

Statisticians have developed a variety of experimental design methods for different purposes, with the expectation that use of these methods will result in increased yields from experiments, quality improvements, and reduced development time or overall costs. Popular experimental design methods include full factorial designs, fractional factorial designs, block designs, orthogonal arrays, Latin square, supersaturated designs, etc. One relatively new design method is called Unifrom Design (UD). Since it was proposed by Fang and Wang in the 1980s (Fang, 1980; Fang et al., 2000; Wang & Fang, 1981), UD has been successfully used in various fields, such as chemistry and chemical engineering, quality and system engineering, computer sciences, survey design, pharmaceuticals, and natural sciences, etc.

Generally speaking, uniform design is a space-filling experimental design that allocates experimental points uniformly scattered in the domain. The fundamental concept of UD is to choose a set of experimental points with the smallest discrepancy among all the possible designs for a given number of factors and experimental runs.

Suppose that there are s factors in an experiment. Without loss of generality we can assume that the experimental domain is the unit cube $C^s = [0,1]^s$ (after making a suitable linear transformation). The aim is to choose a set of n experiment points $P = \{x_1, \ldots, x_n\} \subset C^s$ that is uniformly scattered on C^s. Let M be a measure of uniformity of P such that the smaller M corresponds to better uniformity. Let $Z(n,s)$ be the set of sets of n points on C^s. A set $P^* \in Z(n,s)$ is called a uniform design if it has the minimum M-value over $Z(n,s)$, i.e.,

$$M(P^*) = \min_{P \in Z(n,s)} M(P).$$

Many different measures of uniformity have been defined. However, the centered L_2-discrepancy (CD_2) is regarded as one of the most commonly used measures in constructing the UD tables, the reason is that the CD_2 considers the uniformity not only of P over C^s, but also of all the projection uniformity of P over C^u which is the u-dimensional unit cube involving the coordinates in u, P_u is the projection of P on C^u. Hickernell gave an analytical expression of CD_2 as follows (Fang & Lin, 2003):

$$CD_2(P) = \left[\left(\frac{13}{12} \right)^s - \frac{2}{n} \sum_{k=1}^{n} \prod_{j=1}^{s} \left(1 + \frac{1}{2}|x_{kj} - 0.5| - \frac{1}{2}|x_{kj} - 0.5|^2 \right) \right.$$

$$\left. + \frac{1}{n^2} \sum_{k=1}^{n} \sum_{j=1}^{n} \prod_{i=1}^{s} \left(1 + \frac{1}{2}|x_{ki} - 0.5| + \frac{1}{2}|x_{ji} - 0.5| - \frac{1}{2}|x_{ki} - x_{ji}| \right) \right]^{\frac{1}{2}}$$

where $x_k=(x_{k1},...,x_{ks})$ is the k-th experimental point, s is the number of factors in an experiment, n is the number of runs.

One of the most noteworthy advantages of the uniform design is that it allows an experiment strategy to be conducted in a relatively small number of runs. It is very useful when the levels of the factors are large, especially in some situations in which the number of runs is strictly limited to circumstances when factorial designs and orthogonal arrays can not be realized in practice.

Given that the strength of uniform design is that it provides a series of uniformly scattered experiment points over the domain, this homogeneity in two factors has physically become the spatial uniformity of sampling from a pavement section in x and y directions. The application of uniform design resulted in the generation of a sampling scheme with a UD table consisting of pairs of (x, y) coordinates.

3.2 Fundamental statistics

If $\bar{x}$ is the sample mean of a random sample of size n from a normal population, $X \sim N(\mu, \sigma^2)$, then $Z = (\bar{x} - \mu) / (\sigma / \sqrt{n})$ has a standard normal distribution. A $100(1-\alpha)\%$ confidence interval (CI) can be defined as (Figure 7a), $p\left(-z_{\alpha/2} \leq \dfrac{\bar{x} - \mu}{\sigma/\sqrt{n}} \leq z_{\alpha/2}\right) = 1 - \alpha$.

Hence, If μ and σ are specified, a $100(1-\alpha)\%$ confidence interval on $\bar{x}$ can be then given by

$$\mu - z_{\alpha/2} \cdot \frac{\sigma}{\sqrt{n}} \leq \bar{x} \leq \mu + z_{\alpha/2} \cdot \frac{\sigma}{\sqrt{n}} \tag{3}$$

It can be assumed that the error $E = |\bar{x} - \mu|$ is equivalent to $z_{\alpha/2} \cdot \dfrac{\sigma}{\sqrt{n}}$ (Figure 7b). Then the required sample size will be

$$n = \left(\frac{z_{\alpha/2} \cdot \sigma}{E}\right)^2 \tag{4}$$

That is to say, if $\bar{x}$ is used as an estimate of μ, we can be $100(1-\alpha)\%$ confident that the error $|\bar{x} - \mu|$ will not exceed a specified amount E when the sample size is $n = (z_{\alpha/2} \cdot \sigma / E)^2$ (Montgomery & Runger, 2010). If the specified error level is selected as the fraction of standard deviation of $N(\mu, \sigma^2)$ distribution, i.e., $E = |\bar{x} - \mu| = k\sigma$, where $k \geq 0$, then the Equation 4 can be simplified as $n = (z_{\alpha/2} \cdot \sigma / E)^2 = (z_{\alpha/2} / k)^2$. It should be noted that $z_{\alpha/2} = 1.9600$ if $\alpha = 0.05$; $z_{\alpha/2} = 1.6449$ if $\alpha = 0.10$.

The same argument of sample mean can also be applied to sample standard deviation s. Let X_1, X_2, ..., X_n be a random sample of size n from a normal distribution $N(\mu, \sigma^2)$, and let s^2 be the sample variance. Then the random variable $X^2 = \dfrac{(n-1)s^2}{\sigma^2}$ has a chi-square (χ^2) distribution with $n-1$ degrees of freedom. As shown in Figure 7c, we may write

$$p\left(\chi_{\alpha/2,n-1}^2 \leq \frac{(n-1)s^2}{\sigma^2} \leq \chi_{1-\alpha/2,n-1}^2\right) = 1 - \alpha$$

k $\left(\begin{matrix} E = \|\bar{x} - \mu\| \\ = k\sigma \end{matrix}\right)$	Sample[1] Size	Mean Two-Sided[2] $\dfrac{z_{\alpha/2}}{\sqrt{n}}$	Standard Deviation Two-Sided[3] $\sqrt{\dfrac{\chi^2_{\alpha/2,n-1}}{n-1}}$	$\sqrt{\dfrac{\chi^2_{1-\alpha/2,n-1}}{n-1}}$	One-Sided[4] $\sqrt{\dfrac{\chi^2_{1-\alpha,n-1}}{n-1}}$
1.0	4	0.9800	0.2682	1.7653	1.6140
0.9	5	0.8765	0.3480	1.6691	1.5401
0.8	7	0.7408	0.4541	1.5518	1.4487
0.7	8	0.6930	0.4913	1.5125	1.4176
0.6	11	0.5910	0.5698	1.4312	1.3530
0.5	16	0.4900	0.6461	1.3537	1.2909
0.4	25	0.3920	0.7188	1.2807	1.2318
0.3	43	0.2989	0.7868	1.2128	1.1764
0.2	97	0.1990	0.8587	1.1411	1.1174
0.1	385	0.0999	0.9293	1.0707	1.0591
0.62	10	0.6198	0.5478	1.4538	1.3711
0.44	20	0.4383	0.6847	1.3149	1.2596
0.36	30	0.3578	0.7439	1.2556	1.2114
0.31	40	0.3099	0.7788	1.2208	1.1829
0.28	50	0.2772	0.8025	1.1971	1.1636
0.25	60	0.2530	0.8199	1.1798	1.1493

Note:

Sample size is calculated by $n = \left(z_{\alpha/2} \cdot \sigma / E \right)^2 = \left(z_{\alpha/2} / k \right)^2$.

The two-sided 100(1-α)% confidence interval of sample mean is calculated by

$$\mu - z_{\alpha/2} \cdot \sigma / \sqrt{n} \le \bar{x} \le \mu + z_{\alpha/2} \cdot \sigma / \sqrt{n}.$$

The two-sided 100(1-α)% confidence interval of sample standard deviation is calculated by

$$\sqrt{\chi^2_{\alpha/2,n-1} / (n-1)} \cdot \sigma \le s \le \sqrt{\chi^2_{1-\alpha/2,n-1} / (n-1)} \cdot \sigma.$$

The one-sided 100(1-α)% confidence interval of sample standard deviation is calculated by

$$s \le \sqrt{\chi^2_{1-\alpha,n-1} / (n-1)} \cdot \sigma.$$

Table 3. Summary of 95% confidence intervals of sample mean and sample standard deviation at various error levels and sample sizes for a $N(0, 1)$ distribution.

If s^2 is the sample variance from a random sample of n observations from a normal distribution with known or specified variance σ^2, then a two-sided $100(1-\alpha)\%$ confidence interval on s is

$$\sqrt{\frac{\chi^2_{\alpha/2,n-1}}{n-1}} \cdot \sigma \le s \le \sqrt{\frac{\chi^2_{1-\alpha/2,n-1}}{n-1}} \cdot \sigma$$

As for the one-sided $100(1-\alpha)\%$ upper confidence bound as shown in Figure 7d, we may write

$$p\left(\frac{(n-1)s^2}{\sigma^2} \le \chi^2_{1-\alpha,n-1} \right) = 1-\alpha$$

then the confidence upper bound on s is

$$s \leq \sqrt{\frac{\chi^2_{1-\alpha,n-1}}{n-1}} \cdot \sigma \qquad (5)$$

Table 3 summarizes the 95% confidence interval of sample mean and sample standard deviation at various error levels and sample sizes. Notice that the sample size listed in Table 3 was rounded to its ceiling value.

Figure 8a plots the sample size versus the specified error ($E = |\bar{x} - \mu|$) in terms of standard error (σ) with 95% confidence interval. The two-sided 95% confidence interval on the sample mean and the one-sided 95% upper confidence bound on the sample standard deviation of a $N(0, 1)$ distribution, as a function of sample size, can be illustrated as shown in Figures 8b and 8c, respectively.

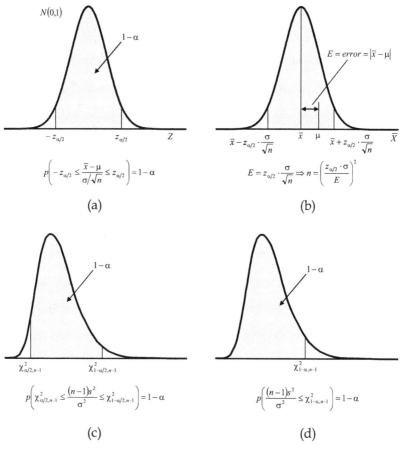

Fig. 7. (a) $100(1-\alpha)$% confidence interval of $N(0, 1)$ distribution, (b) sample size determination with a specified error level, (c) $100(1-\alpha)$% two-sided confidence interval of χ^2 distribution, and (d) $100(1-\alpha)$% one-sided confidence interval of χ^2 distribution.

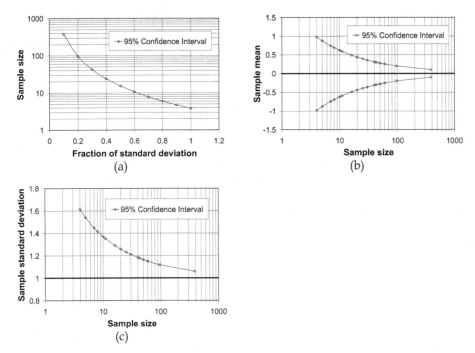

Fig. 8. (a) Sample size versus fraction of standard deviation, (b) 95% two-sided confidence interval of sample mean, and (c) 95% one-sided upper confidence bound of sample standard deviation.

3.3 Sampling scheme and size simulation

In this approach, it was assumed that the air-void contents on a project can be represented by a standard normal $N(0, 1)$ distribution. The data from the $N(0, 1)$ distribution were used to generate five data patterns: random pattern, central segregation pattern, bilateral segregation pattern, central-bilateral segregation pattern, and block segregation pattern (Figure 9). The reasons for selecting these pattern types are as follows:

1. Random pattern: non-segregation, with ideal construction quality.
2. Central segregation pattern: the gap between two augers of an asphalt paver makes coarse aggregate concentrated near the center of the paved area.
3. Bilateral segregation pattern: the gap between the auger and the lateral board of the asphalt paver makes coarse aggregate concentrated near the bilateral regions of the paved area, or provides less compaction of the side area.
4. Central-bilateral segregation pattern: a combined situation of patterns 2 and 3.
5. Block segregation pattern: as demonstrated in gradation segregation, temperature segregation, uneven compaction, etc.

The segregation horizontal strips as shown in Figures 9b, 9c, and 9d were randomly generated using the data in the shaded area of the $N(0, 1)$ distribution, which represent higher air-void contents. In the block segregation pattern (Figure 9e), the $N(0, 1)$ distribution was divided into 6 intervals and the data of each interval were randomly distributed into blocks of pavement sections.

The prospective road section was divided into $n(X)$ (x-direction) × $n(Y)$ (y-direction) cells. The $n(X)$ represents the number of intervals in the x-direction. N points were then assigned to these $n(X)$ × $n(Y)$ cells. Hence, a sampling scheme was defined by $n(X)$, $n(Y)$, and N. For instance, $x30y6n30$ represents 30 runs that were assigned to 30 cells of the 30 × 6 cells based on the UD table. The sampling schemes considered in this study were combinations of various numbers of $n(X)$ and $n(Y)$ —that is, $n(X)$ = 3, 5, 10, 15, 20, 25, 30, 35, 40, 45, 55, 60 and $n(Y)$ = 1, 2, 3, 4, 6—and $N = n(X)$; however, the cases with $n(Y) > n(X)$ were excluded, resulting in a total of 62 cases. Each case was assigned a UD table with minimum CD_2 value. Figures 10a through 10c respectively illustrate the example sampling schemes (i.e., UD tables), $x10y6n10$, $x30y6n30$, and $x60y6n60$, from the uniform design. These sampling schemes are on the same scales of a 900 ft × 24 ft (274 m × 7.32 m) pavement section. The black rectangle cell physically represents the area of which one measure should be sampled randomly.

For this sampling simulation, a total of 2700×72 points with a standard normal distribution of air-void contents were used to generate five data patterns with the following assumptions:

1. Lane width: 24 ft (7.32 m).
2. Time frame of construction: 1 hour with 900 ft (274 m) of HMA placed, i.e., paver speed = 15 ft/min. (4.57 m/min.).
3. One air-void sample is represented by a 4 × 4-in. (10 × 10-cm) square with each square assigned a normalized air-void value.

Each type of sampling scheme per data pattern was simulated 200 times. For each simulation, the sample mean and sample standard deviation were calculated. It should be noted that the data of each simulation were randomly drawn from the cells specified in the UD table with replacement. Consequently, the distributions of the sample mean and standard deviation were generated after 200 simulations. The boxplot was then utilized to characterize the location and dispersion of sample means and standard deviations.

The boxplot illustrates a measure of location (the median [solid black dot or white strip]), a measure of dispersion (the interquartile range IQR [lower quartile: left or bottom-edge of box; upper quartile: right or top-edge of box]), and the possible outliers (data points with light circle or horizontal line outside the 1.5 IQR distance from the edges of box; the most extreme data points within 1.5 IQR distance are marked with square brackets) and also gives an indication of the symmetry or skewness of the distribution.

The Trellis graph introduced by Cleveland in 1993 (Cleveland, 1993) is a graphical way of examining high-dimensional data structure by means of conditional one-, two-, and three-dimensional graphs. As an example, we would like to determine how the sample mean distribution depends on $n(X)$, $n(Y)$, and the data pattern. To inspect this graphically, the simulation results can be split up into groups and can be plot separately as opposed to blurring the effects in a single graph. The Trellis graph of boxplots presented in Figures 11 and 12 was arranged in such a way that each panel consists of all the $n(Y)$ = 1, 2, 3, 4, 6 cases (i.e., 5 boxplots in each panel), each row is made by all the N = $n(X)$ = 3, 5, 10, 15, 20, 25, 30, 35, 40, 45, 55, 60 cases (i.e., 13 panels in a row) with the same data pattern, and each column has 5 panels (i.e., 5 data patterns) with the same $n(X)$. Thus, for each individual column, the effects of data pattern and $n(Y)$ can be examined at the specified $n(X)$; for each individual row, the effects of $n(X)$ and $n(Y)$ can be inspected at the specified data pattern. The Trellis graph was categorized by $n(X)$, $n(Y)$, and five data patterns.

Application of Sampling Strategies for Hot-Mix Asphalt Infrastructure: Quality Control-Quality Assurance Sampling; Specification for Performance Test Requirements

141

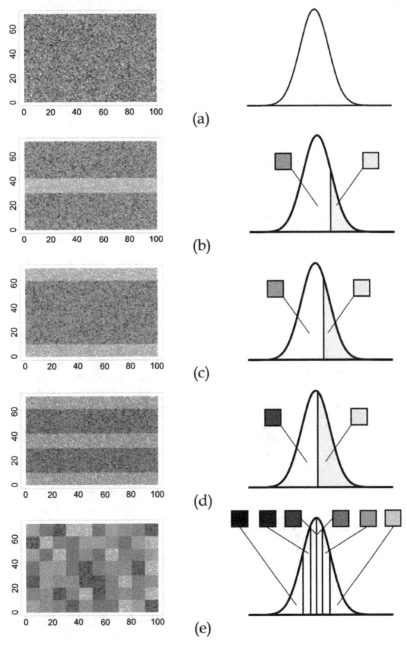

Fig. 9. Schematic illustration of five data patterns: (a) random pattern, (b) central segregation pattern, (c) bilateral segregation pattern, (d) central-bilateral segregation pattern, and (e) block segregation pattern.

The Trellis graphs of the boxplots shown in Figures 11 and 12 summarize respectively the simulation results of sample means and sample standard deviations. Several observations from the Trellis graphs can be made:

1. As $n(X)$ increases, i.e., increase of N, the variations of sample mean and standard deviation reduce exponentially regardless of data patterns.

2. For the segregation data patterns 2, 3, and 4, the increase of $n(Y)$ does benefit the decrease of variation per $n(X)$ and per data pattern. However, no apparent decrease of variation on random or block segregation patterns was perceived. This implies that the UD table provides a uniform sampling strategy. From the perspective of practice, it is suggested that $n(Y)$ should be as large as possible to include all the possible data patterns.

3. It should be noted that the distributions of sample standard deviation at small $n(X)$s exhibit unsymmetrical and skewed distributions due to the intrinsic properties of χ^2 (chi-square) distribution. From the point of view of HMA construction, the $100(1-\alpha)\%$ one-sided upper bound is suggested for the judgment of sample standard deviation. As a result, the smaller the sample standard deviation, the more uniform the construction quality of HMA. Also, from Figures 8b, 8c, 11, and 12, it is apparent that the change of variation decreases sharply at the beginning and the rate of change of variation stabilizes after $N = 20 \sim 30$.

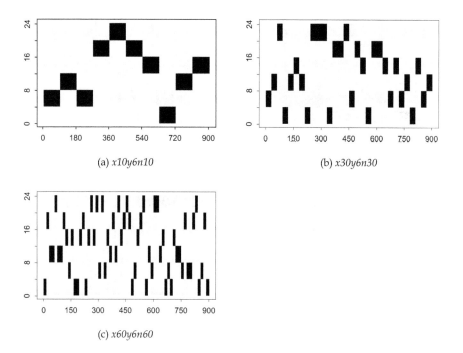

(a) $x10y6n10$

(b) $x30y6n30$

(c) $x60y6n60$

Fig. 10. Examples of UD tables (a) $x10y6n10$, (b) $x30y6n30$, and (c) $x60y6n60$.

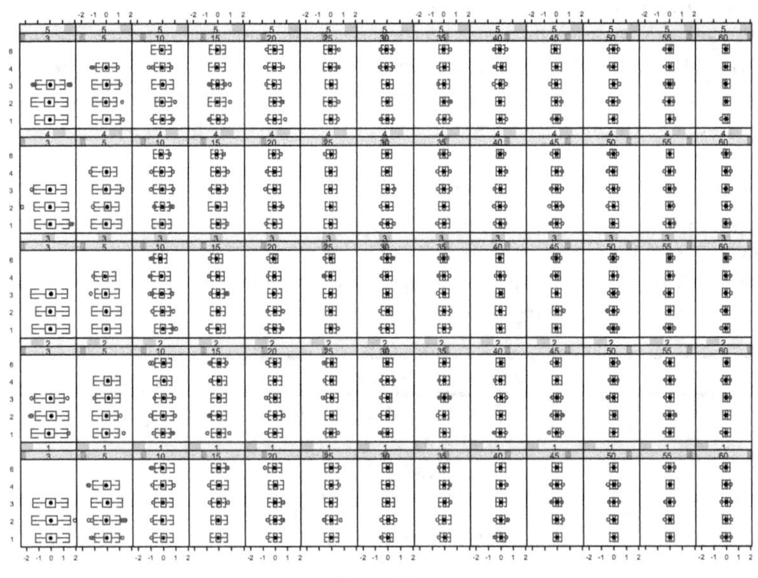

Fig. 11. Trellis graph of boxplots of sample mean categorized by $n(X)$, $n(Y)$, and five data patterns with $N = n(X)$.

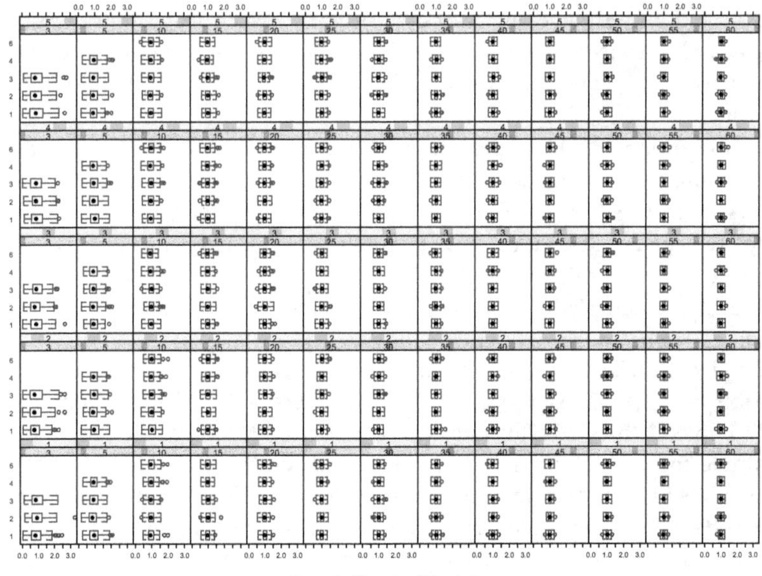

Fig. 12. Trellis graph of boxplots of sample standard deviation categorized by $n(X)$, $n(Y)$, and five data patterns with $N = n(X)$.

3.4 UD demonstration example using two field sections

In this demonstration example, the percent air-void content data of two field pavement sections each with 164 ft (50 m) in length and 36 ft (11 m) in width were acquired by the Pavement Quality Indicator (PQI), which is a non-nuclear density measurement device calibrated with core samples. The percent air-void content was taken by a 3.3 × 3.3-ft (1 × 1-m) square. These two pavement sections served as the "testing sections" of which the paving operation, compaction pattern/effort, and other construction details were verified and corrected (if necessary) by the contractor. Several performance tests were comprehensively conducted by the agency to guarantee that the pavement quality of the whole project met the specifications afterwards. The material properties of two pavement sections, AC-13 and AC-20, are as follows.

	Pavement Section AC-13	Pavement Section AC-20
Asphalt mix type	Asphalt concrete with nominal maximum aggregate size (NMAS) 13 mm	Asphalt concrete with NMAS 20 mm
Binder type	SBS modified binder (equivalent to PG76-22)	
Aggregate type	Diabase (fully crushed, hard rock)	Granite (fully crushed)
Design binder content	5.6%	4.8%
Target air-void content	$N(\mu, \sigma^2) = N(5, 1)$: mean 5%; standard deviation 1%.	
Acceptable air-void content range	5 ± 2 %, i.e., $P(3 \leq AV \leq 7) = 0.95$ of a $N(5, 1)$ distribution	

The measured percent air-void contents are illustrated in Figures 13a and 13b respectively for the AC-13 and AC-20 pavement sections. As can be seen from the figures, the AC-13 section presents high air-void content on the section edges and seems to have a wide variation of air-void content. The AC-20 section appears to have more uniform distribution of air-void content.

To illustrate the proposed QA approach, it was decided that 20 points (20 runs) will be sampled to ensure that the agency is 95% confident that the error $|\bar{x} - \mu|$ will not exceed 0.44σ, i.e., 0.44 percent (Table 3). Two UD tables (Figures 13c and 13d) were generated for both sections which are subdivided into 10 (x-direction) by 11 (y-direction), i.e., x10y11n20. In this case study, the sampling for each UD table was conducted only once. Figures 13e and 13f summarized the sampled, measured, and specified distributions of air-void content. Several findings can be addressed in the following:

1. The sampled distribution based on the UD table matches the measured distribution reasonably well: AC-13 sampled $N(6.29, 1.40^2)$ versus AC-13 measured $N(6.18, 1.43^2)$; AC-20 sampled $N(5.41, 1.22^2)$ versus AC-20 measured $N(5.12, 1.24^2)$.

2. The sample mean, 6.29, of AC-13 section is outside the 95% CI (4.56, 5.44) (Table 3); therefore, it is identified as an "inaccurate" distribution. The sample standard deviation, 1.40 exceeds the 95% one-sided upper bound 1.26 (Table 3); thus, it is designated as an "imprecise" distribution. As a result, the construction quality of AC-13 section is not acceptable because of its "inaccurate" and "imprecise" distribution.

3. On the contrary, the construction quality of AC-20 section is not rejected because of its "accurate" and "precise" distribution: the sample mean 5.41 lies in the 95% CI although

on the high side; the sample standard deviation 1.22 is a slightly less than the 95% one-sided upper bound 1.26.

Accordingly, a proposed QA sampling guideline can be provided by the agency in the following steps:

1. Specify an error level (E) of sample mean in terms of standard deviation of the specified distribution $N\left(\mu,\sigma^2\right)$, i.e., $E=|\bar{x}-\mu|=k\sigma$.

2. Specify the α value to construct a $100(1-\alpha)\%$ confidence interval.

3. Determine the sample size based on the Equation 4, $n=\left(\dfrac{z_{\alpha/2}\cdot\sigma}{E}\right)^2$.

4. Generate a uniform design table (UD table) as the sampling scheme, the X factor should have n (sample size) levels, i.e., $N = n(X) = n$. It is suggested the Y factor should have at least 3 levels per lane, i.e., $n(Y) \geq 3$ per lane.

5. Randomly take the measurement from each (x, y) cell specified in UD table.

6. Check the sample mean $\bar{x}$ from n observations.   If $\mu-\dfrac{z_{\alpha/2}}{\sqrt{n}}\cdot\sigma\leq\bar{x}\leq\mu+\dfrac{z_{\alpha/2}}{\sqrt{n}}\cdot\sigma$

   (Equation 3), then we accept the sample mean; otherwise, the sample mean has been rejected and the agency has to reject the project.

7. Check the sample standard deviation s if step 6 has been satisfied. If $s\leq\sqrt{\dfrac{\chi^2_{1-\alpha,n-1}}{n-1}}\cdot\sigma$

   (Equation 5), then we accept the sample standard deviation; otherwise the project should be rejected because of non-uniformity of construction quality.

4. Why is it inappropriate to take only three samples?

It is not uncommon for agencies to base QA on three samples. However, the following discussion using t distribution is presented showing why it is inappropriate to take only this number of samples for quality assurance. When estimating the mean of a normally distributed population with unknown mean μ and unknown variance σ^2, the t distribution should be applied especially with small sample size. Let $X_1, X_2, ..., X_n$ be a random sample from a normal distribution with unknown mean μ and unknown variance σ^2. The random variable $T=\dfrac{\bar{X}-\mu}{S/\sqrt{n}}$ has a t_{n-1} distribution with $n-1$ degrees of freedom, where $\bar{X}$ is the sample mean and S the sample standard deviation. Now if $\mu=\mu_0$, then

$$T_0 = \frac{\bar{X}-\mu_0}{S/\sqrt{n}}=\frac{\bar{X}-\mu_0-\delta+\delta}{S/\sqrt{n}}=\frac{\bar{X}-(\mu_0+\delta)}{S/\sqrt{n}}+\frac{\delta\sqrt{n}}{S}$$

If $\delta=0$, then it is called the central t_{n-1} distribution. When the true value of the mean is $\mu=\mu_0+\delta$, the distribution of T_0 is termed the noncentral t_{n-1} distribution with the noncentrality parameter $\delta\sqrt{n}/S$. Based on the definition of type II error (β) : $P\{$fail to reject H_0 | H_0 is false$\}$ under the hypothesis testing $H_0:\mu=\mu_0$, $H_1:\mu<\mu_0$, the Type II error is

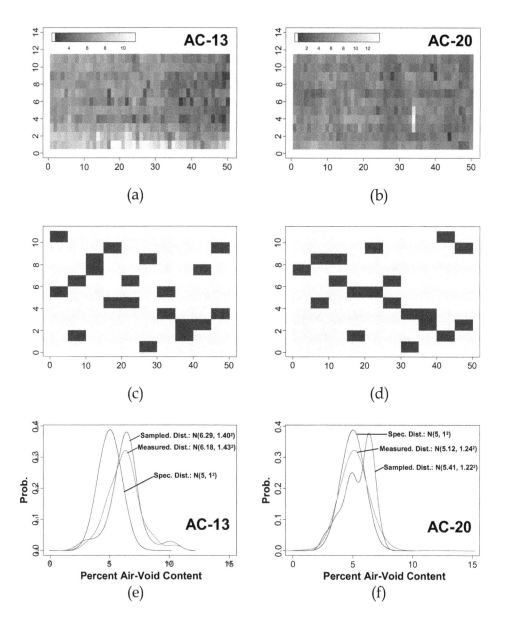

Fig. 13. Image plots of air-void measures for sections (a) AC-13 and (b) AC-20; UD tables *x10y11n20* for sections (c) AC-13 and (d) AC-20; the specification, measured, and sampled distributions for sections (e) AC-13 and (f) AC-20

made only if $T_0 \geq -t_{\alpha,n-1}$ [shown in Figures 14a and 14b respectively for the situations that
(a) $\delta > 0$ and (b) $\delta < 0$] where T_0 is noncentral t_{n-1} distribution. Hence, we have the
probability of Type II error

$$\beta = 1 - T_{n-1}\left(-t_{\alpha,n-1} - \frac{\delta\sqrt{n}}{\sigma}\right), \text{ i.e., } power = 1 - \beta = T_{n-1}\left(-t_{\alpha,n-1} - \frac{\delta\sqrt{n}}{\sigma}\right), \tag{6}$$

where T_{n-1} is the distribution function with $n-1$ degrees of freedom. From Figure 14, it is
apparent that the more positive δ value the larger the β value, i.e., the smaller the power; on
the contrary, the more negative δ value the smaller the β value, i.e., the larger the power.

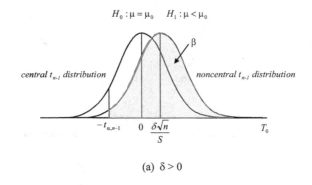

(a) $\delta > 0$

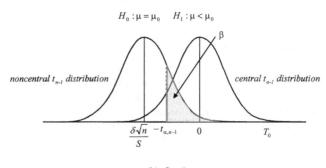

(b) $\delta < 0$

Fig. 14. Definition of type II error (β) of a t distribution under the hypothesis testing:
$H_0 : \mu = \mu_0$, $H_1 : \mu < \mu_0$ for the situations that (a) $\delta > 0$ and (b) $\delta < 0$.

Equation 6 indicates that power is a function of α, n, and δ/S. Figure 15 plots power versus
δ/S at various sample sizes. Under the hypothesis testing $H_0 : \mu = \mu_0$, $H_1 : \mu < \mu_0$, $\alpha = 0.05$
and $n = 3$, the interpretation of Figure 15 is that one will have power greater than 0.8 to
reject the null hypothesis if $\delta/S \leq -2.30$; on the other hand, if $\delta/S \geq -2.30$, then the agency has
insufficient power to reject the null hypothesis that $H_0 : \mu = \mu_0$. It should be noted that, to
increase sample size from three to five, the agency will have power greater than 0.8 if $\delta/S \leq$
-1.37; that is, the agency can detect smaller mean difference from $2.30S$ down to $1.37S$ by

increasing two samples. In sum, by taking only three samples out of a project, the agency will have insufficient power to reject $H_0 : \mu = \mu_0$ given that H_0 is false unless the quality of the project delivered by the contractor is so poor that the agency is confident enough to reject the project.

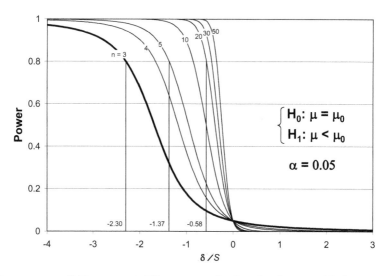

Fig. 15. Power versus δ/S curves at different sample sizes for the one-sided t-test at a significance level $\alpha = 0.05$.

5. Findings and conclusions

For the Case I study, an attempt has been made to illustrate an approach and the extent of testing required using a performance test to insure reasonable quality in as-placed HMA. Stabilometer S-value test results were used in this example since extensive data were available. It should be emphasized that the same approach could be applied using other test parameters to control the quality of the as-constructed mix.

Based on stabilometer test results, the brief discussion of hypothesis testing, and the simulation results of sampling scheme and size, the following observations and suggestions are offered:

1. Cooperation between the agency and the contractor is essential. It is necessary to have the testing process, test equipment, test results, and specimen preparation as consistent as possible between the two organizations.

2. The sampling simulation of the Case I demonstration example suggests that the sample size required to stabilize the sampling consistency and sampling stabilization is around 50 ~ 70 for the placement of 15,000 tons HMA.

3. Likely, sampling as noted (2) is perhaps impractical. However, increasing the sample size is actually beneficial for both the agency and the contractor since it reduces the potential for dispute and guarantees the quality of the constructed mix. By extension, it is advisable for the agency to provide incentives to encourage the contractor to increase sampling size and testing.

4. To ensure the success of the proposed QC/QA guidelines, the contractor's minimum value of the testing null hypothesis must exceed that required by the agency.

5. From the Caltrans case study, the μ_{min} criterion depended not only on the contractor's α value and the agency's power level as expected but also on the k value that the agency would select for use. The μ_{min} criterion can be smaller if both the agency and the contractor require low power level and high α level and/or the agency increases the k value.

A concluding general observation relates to the concern for developing longer lasting pavement at this period of time because of increased costs of both pavement materials and increased traffic that must be accommodated. The added costs of testing by both the contractor and the agency are a very small proportion of the total costs associated with long lasting pavements. Accordingly an "attitude adjustment" for both parties relative to QC and QA testing would enhance long-term pavement performance.

From above discussion of Case II for determining sample size, simulation results of the sampling size and sampling scheme using UD tables, along with a demonstration example, the following observations and suggestions are offered:

1. It is important to recognize that the agency can be $100(1-\alpha)\%$ confident that the error $|\overline{x} - \mu|$ will not exceed a specified amount E if and only if the sample size is $n = \left(z_{\alpha/2} \cdot \sigma / E\right)^2$. The variations of sample mean and sample standard deviation for the 900 ft HMA paving simulation (Figures 8, 11, and 12) suggests that the minimum sample size required to stabilize the variation is around 20 ~30.

2. The UD table not only provides the most representative sampling scheme with the sample size for a given specified error level by the agency but also minimizes the possible effect of the underlying data pattern. Moreover, the UD table gives the agency a more unbiased "random" sampling scheme that can be followed in the quality assurance process.

3. The sample mean and sample standard deviation criteria proposed in the QA guideline demonstrates the accurate/inaccurate and precise/imprecise concept of sampling outcomes. If the sample mean is located in the range of $100(1-\alpha)\%$ confidence interval, then it is accurate. Precision is a term to describe the degree of data dispersion; if the sample standard deviation is less than the $100(1-\alpha)\%$ one-sided upper bound, then it is precise. The case study presents a very good example of an inaccurate/ imprecise case of the AC-13 field section and an accurate/precise case of the AC-20 field section. The quality of a project can only be accepted if and only if these criteria have been fulfilled simultaneously.

4. The proposed QA guideline with the introduction of the UD table is relatively simple, practical, and robust. The sample mean and sample standard deviation criteria are rational enough for both the agency and the contractor to agree upon.

5. It should be emphasized that the proposed QA approach could be applied with other performance measurement parameters to control the quality of the as-constructed mix, such as thickness, stabilometer testing as used in California, performance testing of fatigue and rutting, etc. Moreover, the decision-making based on this proposed QA approach can also be a basis for pay factor determination.

6. By taking only three samples out of a project, the agency will have insufficient power to reject $H_0 : \mu = \mu_0$ given that H_0 is false unless the quality of the project delivered by the

contractor is so poor that the angency is confident enough to reject the project. However, by increasing sample size from three to five, the agency can detect smaller mean difference from 2.30S down to 1.37S by simply increasing two samples.

7.   It is likely that the proposed sampling size is impractical. In this regard, the alternative is to establish a "testing section" similar to those in the case study and follow the proposed QA approach with the minimum sampling size (at least greater than 20) to ensure that the compaction pattern/effort, paving operation, and other construction details are appropriate to guarantee that the pavement quality meets the specifications.

6. Acknowledgments

The research associated with the first case study was conducted as a part of the Partnered Pavement Research Program supported by the California Department of Transportation (Caltrans) Division of Research and Innovation. Special thanks go to Mr. Kee Foo of Caltrans who provided the stability data from Caltrans projects. The contents of this paper reflect the views of the authors who are responsible for the facts and accuracy of the information presented and do not reflect the official views of the policies of the State of California or the Federal Highway Administration.

The field data associated with the second case study was sponsored by the Ministry of Transport of the People's Republic of China. The contents of this paper reflect the views of the authors who are responsible for the facts and accuracy of the information presented and do not reflect the official views of the policies of the Ministry of Transport of the People's Republic of China.

7. References

California Department of Transportation (Nov. 2007). *Standard Specifications*, Sacramento, Calif., USA

Cleveland, W.S. (1993). *Visualizing Data*, Hobart Press, ISBN 978-0963488404, Summit, NJ, USA

Fang, K.T. (1980). The Uniform Design: Application of Number Theoretical Methods in Experimental Design. *Acta Mathematicae Applagatae Sinica*, Vol.3, pp. 353-372, ISSN 1618-3832

Fang, K.T.; Lin, D.K.J, Winker, P. & Zhang, Y. (2000). Uniform Design: Theory and Application. *Technometrics*, Vol.42, No.3, pp. 237-248, ISSN 1537-2723

Fang, K.T. & Lin, D.K.J (2003). Uniform Experimental Designs and Their Applications in Industry. In: *Handbook of Statistics 22*, Khattree, R. & Rao, C.R., pp. 131-170, ISBN 0444-506144

Montgomery, D.C. & Runger, G.C. (2010). *Applied Statistics and Probability for Engineers*, John Wiley & Sons, Inc., ISBN 978-0-470-05304-1, USA

Stone, C.J.A. (1996). *Course in Probability and Statistics*, Duxbury Press, ISBN 0-534-23328-7, Pacific Grove, Calif., USA

Tsai, B.-W. & Monismith, C.L. (2009). Quality Control – Quality Assurance Sampling Strategies for Hot-Mix Asphalt Construction. *In Transportation Research Board: Journal of the Transportation Research Board*, No.2098, pp. 51-62, ISSN 0361-1981

Wang, Y. & Fang, K.T. (1981). A Note on Uniform Distribution and Experimental Design. *KeXue TongBao*, Vol.26, pp. 485-489, ISSN 0250-7862

Part 3

Quality Control for Biotechnology

8

Genomic Microarray Quality Assurance

Catherine D. Kashork, Lisa G. Shaffer and Kyle S. Sundin
Signature Genomics Laboratories
USA

1. Introduction

The use of microarray technology is revolutionizing the field of clinical cytogenetics. This new technology has transformed the cytogenetics laboratory by the adaptation of techniques that had previously been the province of molecular geneticists. Proficiency with these techniques is now a must for the modern cytogeneticist. This chapter will focus on quality assurance principles associated with microarray analysis for the diagnosis of copy number changes associated with genetic disease.

Microarrays consist of a glass slide or other solid support on which small amounts of DNA ("probes" or "targets") are deposited and immobilized in an ordered fashion (DeRisi et al., 1996; Schena et al., 1995). Probes vary in size from oligonucleotides manufactured to represent genomic regions of interest (25-85 base pairs [bp] of DNA) to large genomic clones such as bacterial artificial chromosomes (BACs, 80-200 thousand base pairs [kb]). Analysis methodology for microarray-based comparative genomic hybridization (aCGH) is consistent regardless of the probe content. First, DNA is extracted from a test sample (e.g., blood, skin, cells from pregnancy). The patient DNA is labeled with a fluorescent dye, while a DNA from a normal control (reference) sample or pooled control samples is labeled with a different-colored fluorescent dye. The two genomic DNAs, test and reference, are then mixed together and applied to the array. Because the DNAs have been denatured, they are single strands; when applied to the array, the single-strand DNAs hybridize with the arrayed single-strand probes. Using a dual-color scanner, digital images are captured, and the relative fluorescence intensities of the hybridized labeled DNA probes are quantified. The fluorescence ratio of the test and reference hybridization signals is determined at different positions along the genome and provides information on the relative copy number of sequences in the test genome compared to the normal diploid genome, enabling the detection of submicroscopic chromosomal deletions and duplications at an unprecedented level (Beaudet & Belmont, 2008; Shaffer & Bejjani, 2009).

Launching a new assay in the clinical setting requires an effective validation of the assay, clear protocols for use at the bench and clearly defined quality assurance (QA) and quality control (QC) procedures prior to the launch. Every laboratory must develop a strong Quality Management System (QMS) that is coordinated with the defined policies under regulatory bodies, such as CLIA '88 (Schwartz, 1999), College of American Pathologists and state regulating agencies. These agencies perform rigorous inspections and verify that a diagnostic laboratory follows defined principles to ensure quality patient care and correct diagnosis. This chapter covers many of the QA and QC principles identified and monitored for laboratories offering microarray-based diagnostics.

2. Quality systems with strong monitoring for quality metrics

Regulatory bodies require diagnostic laboratories to build a strong QMS (Deming, 2000). A robust QMS integrates the organization's processes, policies and procedures for total quality management. In the diagnostic laboratory industry, CAP and other accrediting bodies require defined metrics throughout all phases of testing, including pre-analytical, analytical and post-analytical.

For microarray technology, pre-analytical metrics may include assessment of DNA quality and yield. Each laboratory must define the ideal quality of DNA prior to implementing the assay into clinical testing. In our experience, a gel assessment that indicates clean genomic DNA free of RNase and degradation should lead to quality microarray results. If a DNA specimen has artifacts or appears to have degraded (Fig. 1), the laboratory should inform the client that results may be compromised because of DNA quality or obtain a new specimen from which to perform the analysis. In addition to a visual assessment of the DNA via gel electrophoresis, the laboratory should assess the DNA yield post-extraction. The quantity of DNA required in the analytic labeling phase of microarray analysis determines DNA yield requirements. Spectrophotometric assessment of DNA offers two indicators of quality DNA, including quantity and purity. A 260/280 nm reading indicates quantity, and a 260/230 nm reading indicates purity. These measurements are imperative for the downstream labeling process. Insufficient DNA quantity and quality (purity) will compromise successful microarray analysis. The spectrophotometer measures optical density (OD), which is the physical process of absorbing light. The OD, or absorbance, is calculated as a mathematical quantity. OD readings for pure DNA should measure at 1.8 (Sambrook & Russell, 2001). Our laboratory uses OD measurements from 1.4 to 1.8, although quality of labeling product can be compromised at the lower OD readings. There are many causes of poor yield, including compromised technique during extraction and poor sample quality (e.g., from increased age or exposure).

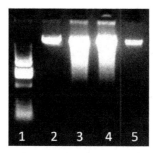

Fig. 1. Gel electrophoresis for the assessment of DNA degradation. Lane 1 has the molecular mass standard. Lanes 2 & 5 show high molecular weight samples that do not exhibit any signs of degradation. The two samples in lanes 3 & 4 show lower molecular weight DNA below the main high molecular weight bands in the other lanes. Degraded DNA typically leads to compromised array results.

Assessment of quality should be implemented throughout all phases of testing including the analytic phase. For aCGH, analytical metrics may include, but are not limited to, spectrophotometric assessment of the labeling product and the identification of labeling efficiency, which has an impact on results. Microarray analysis usually requires a dye incorporation using a random priming method. The dyes are tagged to a promoter molecule, and the promoter molecule is incorporated into the genomic DNA. As with pre-analytic assessment of DNA, the quantity assessment of the labeling product is performed by assessing the 260/280 nm readings from the spectrophotometer. Laboratories should define post-labeling quantity requirements that indicate labeling efficiency. In addition to labeling efficiency as a quality indicator, visual assessment of the set-up of labeling product on the array should be considered. In the event that there are air bubbles or non-complete contact of array product to the hybridization area (Fig. 2), the quality of the microarray result may be compromised and should be documented.

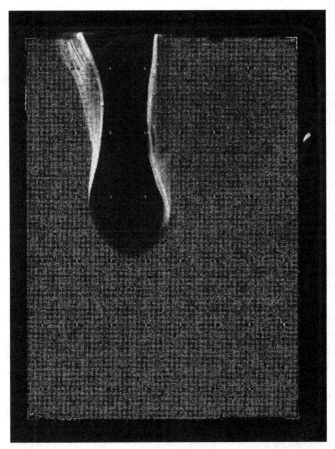

Fig. 2. Microarray after hybridization with an air bubble. The air bubble creates an area of incomplete contact of array product to the hybridization area, which compromises the array.

Post-analytic assessments may include average standard deviation (SD), intensity and background values. The SD value is the standard deviation of the normalized $\log_2$ intensity ratios for autosomal regions (excluding large copy number imbalances) and provides a measure of quality for aCGH experiments (Vermeesch et al., 2005). The SD value provides a quantitative metric that is relative to the overall noise on an array. As the overall noise of an aCGH experiment increases, so does the SD value. Our laboratory has established SD values that indicate whether an aCGH experiment is optimal, suboptimal, or failed. In our laboratory the average SD is used daily to monitor the collective SD values for all patients. Daily monitoring of the average SD value allows for establishment of a system to monitor the average SD over time. If a shift in the average SD value is observed, the laboratory processes can be evaluated to determine the potential cause and potentially prevent a system-wide failure. Each laboratory will need to define and validate a quality metric to measure the quality of the array and implement a system to track the performance of the metric.

In addition to monitoring the SD value of the array, the signal intensity of the two fluorescent dyes relative to the background can be monitored. These two values can be tracked independently or together by monitoring the signal-to-noise ratio (SNR). The SNR is the signal intensity divided by the background noise. Low signal intensity or high background noise will result in a low SNR value. A low SNR is an indicator for poor-quality array data (Basarsky et al., 2000). Low signal intensities can result from several factors, including poor fluorescent dye incorporation in labeling, inadequate denaturation of the probe, inadequate quantity of the probe, and suboptimal hybridization. Several factors can result in high background noise, including labeling reaction impurities, drying of the array during hybridization or during the post-hybridization washes, and incorrect Cot-1 to probe ratio. Constant monitoring of these metrics allows the laboratory staff to anticipate potential system failures leading to failed or inaccurate findings.

3. Verification of array results

In addition to the microarray assessments, there are other post-analytical assessments of quality that validate the microarray findings and lead to a complete, quality result used by the clinician for the diagnosis of the patient. Methods for confirmation of array results may include fluorescence *in situ* hybridization (FISH), multiplex ligation-dependent probe amplification (MLPA), quantitative PCR (Q-PCR) and other PCR techniques.

FISH is an established technique that is used to identify numerical and structural chromosome abnormalities by using fluorescently labeled DNA probes to detect the presence or absence of the DNA in the interphase nucleus or in metaphase, the stage of active cell division when the chromosomes are visibly condensed and can be observed in a microscope (Kashork et al., 2010). FISH commonly uses unique-sequence BAC probes; depending on the specific probe used, the resolution of metaphase FISH is ~80-200 kb (Shaffer et al., 2001). In our experience, deletions are easy to visualize by FISH. However, tandem duplications represent a challenge to any laboratory using confirmatory FISH because the duplicated material is not of sufficient distance from the original genomic location to generate a distinct fluorescent signal to allow detection by interphase or metaphase analysis. In some cases the intensity of the signal may be twice as intense on the duplicated homolog compared to the normal homolog, but this is not always the case.

In addition, although microarray analysis can detect DNA copy number changes, it does not identify the provenance of the abnormality; seemingly identical array results may be caused by distinct molecular mechanisms. Complete understanding of the rearrangement so that accurate genetic counseling can be provided requires visualization of the rearrangement, which can be accomplished with FISH. For example, a copy-number gain identified by microarray analysis (Fig. 3) may be a duplication, an insertion, a marker chromosome or an unbalanced translocation (Fig. 4).

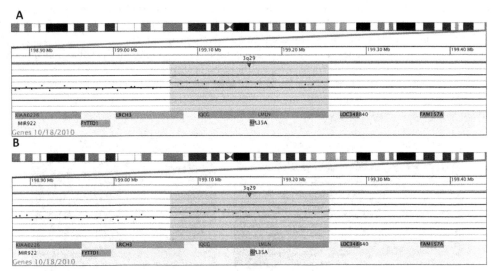

Fig. 3. (A-B) Microarray plots from two subjects showing identical single-copy gains of 17 oligonucleotide probes from the terminal long arm of chromosome 3, approximately 189 kb in size (chr3:199,067,024-199,255,755, hg18 coordinates). Probes are ordered on the x-axis according to physical mapping positions, with the most proximal 3q29 probes to the left and the most distal 3q29 probes to the right. Values along the y-axis represent log_2 ratios of patient:control signal intensities. Results are visualized using Genoglyphix (Signature Genomics, Spokane, WA).

One disadvantage of FISH visualization of microarray results, particularly when using high-density oligonucleotide arrays, is that high-density oligonucleotide arrays can detect abnormalities well below the size of the smallest FISH probes, which are 100-350 kb in size. However, PCR-based methodologies such as MLPA and Q-PCR can visualize small gains and losses. MLPA targets the region of interest with two oligonucleotide probes, one probe containing a forward primer sequence and the second probe containing the reverse primer sequence (Schouten et al., 2002). The oligonucleotide probes are allowed to hybridize to the DNA followed by a ligation step. If the two probes are adjacent to each other the ligation will combine the two probes into a single probe with a fluorescently tagged forward primer on one end and reverse primer on the other end. The probe is amplified by PCR and only the ligated probe is amplified. The amplified product is dependent on the number of target sites present in the DNA. The forward primer is fluorescently labeled, which allows a comparison of the ratio of the fluorescent intensity between reference sample and the test subject to determine the relative quantity of the probe.

Q-PCR amplifies and simultaneously quantifies the relative amount of DNA when compared against a reference. Two Q-PCR methods have been developed. The first method uses fluorescent dyes that intercalate nonspecifically with the double-stranded DNA which produces fluorescent signals relative the quantity of DNA present (VanGuilder et al., 2008). The ratio is compared against a normal reference to confirm the relative quantity of the sample to the control. The second method uses a fluorescently labeled probe that is targeted to the region of interest. The fluorescently labeled probe has a fluorescent reporter and a quencher to hide the fluorescent signal until the region is amplified (Udvardi et al., 2008). During each round of the PCR process, the exonuclease activity of the polymerase releases the fluorescent reporter, unquenching the signal and allowing detection.

Additional PCR-based methods such as polymorphic microsatellite analysis have also been used as a confirmatory assay for aCGH. Although these molecular assays can confirm a copy number gain or loss, they cannot reveal the chromosomal rearrangement or mechanism giving rise to the copy number variant (CNV). Each laboratory must determine the appropriate confirmatory assay to meet its needs.

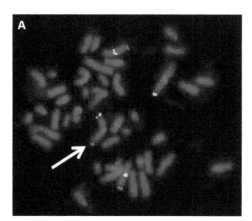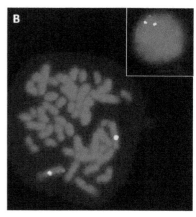

Fig. 4. (A) FISH visualization of the gain shown in Fig. 3A revealed an unbalanced translocation of the 3q29 material to 1p. BAC clone RP11-23M2 from 3q29 is labeled in red, and BAC clones RP11-9A9 from 3q11.2 and RP11-438F14 from 1q44 are labeled in green as controls. The presence of one red signal on one of the chromosome 1 homologues indicates translocation of 3q29 onto 1p (arrow). (B) FISH of the gain shown in Fig. 3B. BAC clone RP11-159K3 from 3q29 is labeled in red, and chromosome 3 centromere probe D3Z1 is labeled in green as a control. The presence of two red signals on metaphase FISH rules out an unbalanced translocation, while the additional finding of three red signals on interphase FISH (inset) suggests a duplication. The patient shown in Fig. 3A may have inherited the unbalanced translocation from a parent with a balanced translocation, whereas the duplication in the patient shown in Fig. 3B may have arisen *de novo*, although parental testing is necessary to confirm the inheritance.

4. Environmental controls

Over time, specific environmental controls have been developed and implemented within the laboratory to ensure quality microarray diagnostics. For aCGH, ozone is an important environmental factor to control. Ozone is a common pollutant found in the lower atmosphere and is the primary component of smog. Ozone is formed when nitric oxides and volatile organic compounds (VOCs) react in the presence of sunlight (US Environmental Protection Agency [EPA], http://www.epa.gov/air/ozonepollution, last accessed May 9, 2008). Nitric oxides and VOCs are emitted by motor vehicle exhaust, industrial emissions, gasoline vapors, chemical solvents, and natural sources. Consequently, ozone levels are higher in urban and industrial areas, especially during the summer months.

National ozone standards have been established by the EPA to protect public health. The established standard peak ozone level set by the EPA is 80 ppb, which is based on the annual fourth maximum 8-hour average (EPA, http://www.epa.gov/air/ozonepollution, last accessed May 9, 2008). The EPA has also established an air quality index system for monitoring the daily pollution levels. The "good" air quality range is 0 to 60 ppb. The fluorescent dyes commonly used in aCGH are sensitive to ozone levels as low as 5 to 10 ppb (Branham et al., 2007; Fare et al., 2003). Thus, ozone levels considered normal for environmental standards are well above those ranges, demonstrating sensitivity of the dyes.

Ozone has been shown to strongly affect dyes that are commonly used in aCGH, including cyanine 5 (Cy5) and Alexa dyes (Alexa Fluor 647 and Alexa Fluor 5) and, to a lesser extent, cyanine 3 (Cy3) and the Alexa equivalents (Alexa Fluor 555 and Alexa Fluor 3) (Branham et al., 2007; Byerly et al., 2009; Fare et al., 2003). Several studies have identified the post-hybridization washes as the most sensitive period for exposure to ozone (Branham et al., 2007; Byerly et al., 2009; Fare et al., 2003). These studies have demonstrated the difficulty with which laboratories identify the source of ozone-related problems, especially considering the extremely low levels of ozone (5 to 10 ppb) that cause these problems, the duration of exposure (as little as 10 to 30 seconds), and the seasonal emergence of ozone itself. The effects of ozone must be addressed when aCGH is performed, particularly in a clinical diagnostic setting, where it is critical to have consistent high quality and reproducible results. Failure to protect the fluorescent dyes from ozone during the post-hybridization washes will result in considerable negative impact on the array data. The implementation of quality control measures such as ozone reduction and monitoring to ensure high-quality aCGH results is mandatory for any aCGH laboratory. There are many commercially available enclosures and scrubbers designed to protect the dyes during the post-hybridizations washes. Some laboratories have gone as far as developing ozone-free rooms where post-hybridization washes and the subsequent scanning and analysis are performed. The latter is most desirable but may not always be feasible.

In addition to ozone degradation, the dyes are also photosensitive and often must be used in a reduced-light environment. Systems should be implemented to prevent photobleaching of the fluorescent dyes. To mitigate against the effect of photobleaching, the dyes should be protected from the light whenever possible. This can be done by using indirect lighting in the work area, using amber tubes, covering the samples with tin foil or placing the samples in areas with little or no light when not being directly handled.

5. Normalization

Normalization, which aims to separate biologically relevant signal from experimental artifacts, is a crucial step of microarray analysis (Neuvial et al., 2006). Each laboratory must identify a system for normalization. Most microarray vendors offer software with built-in normalization methods optimized for their own platforms. Laboratories can use a normalization package that is developed by the microarray vendor or can develop their own package. One normalization method that is used in the laboratory is the locally weighted polynomial regression (LOESS) (Cleveland, 1979). This normalization applies a spatial correction to correct position-dependent non-uniformity of signals across the array.

Another normalization system used in the laboratory is the Qspline fit normalization (Workman et al., 2002). This normalization compensates for the inherent differences in signal intensities between the two dyes. There are many normalization methodologies available; each laboratory will have to define its method of choice.

6. Automation

One of the key elements for any clinical assay is reproducibility. By replacing manual processes with automation, a laboratory can substantially improve the consistency and reproducibility of its daily operations. In addition, automation can increase throughput, which is often an advantage for a growing laboratory and helps to reduce the dependency on staffing levels. The laboratory protocols that have been successfully automated include DNA isolation, labeling and hybridization, washing and analysis.

However, automation can present several challenges. Because it is based on a plate format, if the input materials are flawed or a technical issue occurs (e.g., labeling master mix or an automation failure) it will impact the entire plate, which may consist of 48 or more patients. The consequences of a failure of this magnitude are substantial in terms of cost and the potential loss of the sample.

7. Multiplexing of microarray platforms

As laboratories become more accustomed to using microarrays, the demand for the assay may increase. In addition to automation, multiplexed array formats can help the laboratory satisfy the increased demand. Multiplexed array formats allow for the simultaneous hybridization of 2 to 12 or more samples depending on the probe coverage of the array and the array manufacturer. The multiplex design (Fig. 5) has many positive features including decreased costs and higher throughput. As this technology continues to advance, higher multiplexed formats are likely to be developed.

Although increased throughput has its advantages, it also creates challenges for any quality system. The laboratory must ensure that there is no cross contamination between each sub-array, which would affect patient samples. Some manufacturers have included QA/QC features into the development of multiplexed arrays, such as tracking controls that can be spiked into the experiments that identify unique positions on the array such that each position of the array can have a unique tracking control. Unique tracking controls are added to each sample before they are introduced to the array, which allows the laboratory to monitor each sub-array for cross contamination. In the event of cross contamination or leaking between the sub-arrays, the laboratory can determine the sample involved in the cross-contamination event based on the unique tracking controls involved.

8. Validation

As laboratories begin to adopt microarray technology within their facility, they should identify the validation criteria which they are responsible for meeting. The American College of Medical Genetics (ACMG; Shaffer et al., 2007) and other US state guidelines have been developed to ensure laboratories have thoroughly tested and reviewed the capacity and expectation of the assay prior to clinical release. These validations include testing known abnormal specimens to verify the expected outcomes. Section E13.2 of the ACMG

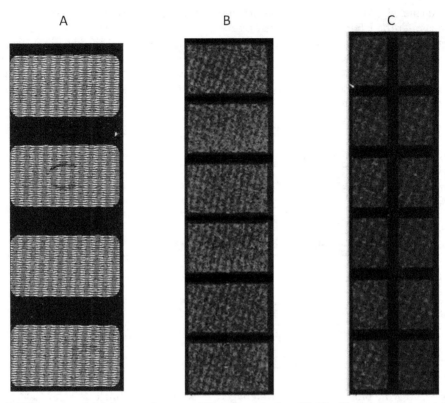

Fig. 5. Different multiplex array formats. (A) Agilent 4-plex, (B) NimbleGen 6-plex, and (C) NimbleGen 12-Plex.

guidelines distinguishes between different levels of validation depending on platform type (e.g., FDA-approved, investigation-use-only/research-use-only, or home-brew microarrays) and requires a demonstration of expertise of array performance and analysis through defined validation criteria for new microarray platforms, new versions of existing microarrays, and new lots of the same microarray. Some states have additional validation requirements. For example, New York has the Clinical Laboratory Evaluation Program (CLEP). Accreditation through CLEP requires additional quality assurances and validation criteria to which laboratories must adhere. Outside of accreditations, laboratories need to consider state regulations when considering offering their testing services nationwide. Some states have regulations that not only impact laboratories that reside in that state but also impact laboratories that test samples from that particular state.

9. Control samples

When defining the control specimen of choice, a laboratory can choose to use same-sex or sex-mismatched controls from the patient. Same-sex controls offer detection of autosomal gains and losses, and complex sex chromosome abnormalities are more easily visualized. Sex-mismatched controls offer the same detection of autosomal gains and losses but can be more challenging when assessing sex chromosome changes. However, sex-mismatched controls offer the laboratory an internal assessment of hybridization success because of the expected deviations when comparing a male against a female. The deviations that are inherent to sex-mismatched controls are a result of the copy number variation of the X/Y chromosome ratio. When a female genome is compared to a male genome, there is an apparent gain of chromosome X (two copies in the female against the single copy in males) and a loss of chromosome Y (Fig. 6).

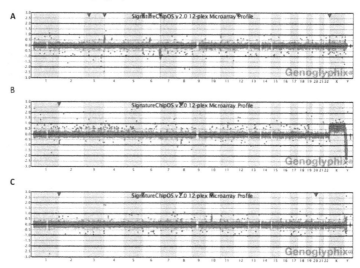

Fig. 6. Microarray plots from three different gender parings: male/male, female/male, and female/female. The probes are ordered on the x-axis according to physical mapping positions starting with chromosome 1 on the left and the X and Y chromosomes on the right. Values along the y-axis represent $\log_2$ ratios of patient:control signal intensities. (A) A same-sex male/male comparison showing identical dosage at the X and Y regions. (B) An opposite-sex female/male comparison showing a gain of the X chromosome (two copies in the female versus a single copy in the male) and a loss of Y (no copies in the female versus one copy in the male). (C) A same-sex female/female comparison showing identical dosage at the X region and no hybridization at the Y region. Results are visualized using Genoglyphix (Signature Genomics, Spokane, WA).

In addition to determining the ideal sex of the control, the laboratory must decide if a single control or pooled controls will be used in the facility. Often, these decisions are made based on available controls. Some institutions have used a consistent male and female control for easy identification of known CNVs and for monitoring the performance of arrays. However, in the absence of consistent candidates for controls, laboratories can create or purchase a pooled DNA control. All variations of these controls must be assessed with any change in the pool so that they will not significantly impact reporting. In addition, CNVs present in the control DNA can be used as a positive indicator of assay performance when using same-sex controls. However, pooled control samples will have diluted CNVs, which may not be apparent on the microarray results or may appear as mosaics or background noise.

10. Conclusion

This chapter highlights many of the quality assurance principles that impact a laboratory setting up or using aCGH. This is not an exhaustive set of challenges to implementation as there may be lab-, region-, environment- and vendor-specific variations. Each laboratory should perform initial quality verification at the time of test development. Post development, a thorough validation must be performed, which may uncover variation that should be controlled prior to launching the clinical assay. A laboratory's role in developing new assays should include an established, documented and maintained quality system that ensures that the test conforms to specified requirements and ultimately leads to accurate results.

11. Acknowledgments

We thank Aaron Theisen (Signature Genomics, Spokane, WA) for his critical review of our manuscript.

12. References

Basarsky T., Verdnik D., Willis D. & Zhai J. (2000). An overview of a DNA microarray scanner: design essentials for an integrated acquisition and analysis platform, In: *Microarray biochip technology*, Schena M. (Ed.). Eaton Publishing, Natick, MA

Beaudet A.L. & Belmont J.W. (2008). Array-based DNA diagnostics: let the revolution begin. *Annu Rev Med*, Vol. 59. pp. 113-129, ISSN 0066-4219

Branham W.S., Melvin C.D., Han T., Desai V.G., Moland C.L., Scully A.T. & Fuscoe J.C. (2007). Elimination of laboratory ozone leads to a dramatic improvement in the reproducibility of microarray gene expression measurements. *BMC Biotechnol*, Vol. 7. pp. 8, ISSN 1472-6750

Byerly S., Sundin K., Raja R., Stanchfield J., Bejjani B.A. & Shaffer L.G. (2009). Effects of ozone exposure during microarray posthybridization washes and scanning. *J Mol Diagn*, Vol. 11. No. 6, (Nov), pp. 590-597, ISSN 1943-7811

Cleveland W.S. (1979). Robust locally weighted regression and smoothing scatterplots. *J Amer Statist Assoc*, Vol. 74. pp. 829-836

Deming W.E. (2000). *The New Economics For Industry, Government & Education* (2nd), The MIT Press, Cambridge, MA

DeRisi J., Penland L., Brown P.O., Bittner M.L., Meltzer P.S., Ray M., Chen Y., Su Y.A. & Trent J.M. (1996). Use of a cDNA microarray to analyse gene expression patterns in human cancer. *Nat Genet*, Vol. 14. No. 4, (Dec), pp. 457-460

Fare T.L., Coffey E.M., Dai H., He Y.D., Kessler D.A., Kilian K.A., Koch J.E., LeProust E., Marton M.J., Meyer M.R., Stoughton R.B., Tokiwa G.Y. & Wang Y. (2003). Effects of atmospheric ozone on microarray data quality. *Anal Chem*, Vol. 75. No. 17, (Sep 1), pp. 4672-4675, ISSN 0003-2700

Kashork C.D., Theisen A. & Shaffer L.G. (2010). Diagnosis of cryptic chromosomal syndromes by fluorescence in situ hybridization (FISH). *Curr Protoc Hum Genet*, Vol. Chapter 8. Unit 8.10 (Oct), pp. 11-20, ISSN 1934-8258

Neuvial P., Hupe P., Brito I., Liva S., Manie E., Brennetot C., Radvanyi F., Aurias A. & Barillot E. (2006). Spatial normalization of array-CGH data. *BMC Bioinformatics*, Vol. 7. pp. 264, ISSN 1471-2105

Sambrook J. & Russell D. (2001). *Molecular Cloning: A Laboratory Manual* (3rd), Cold Spring Harbor Laboratory Press, Woodbury, NY

Schena M., Shalon D., Davis R.W. & Brown P.O. (1995). Quantitative monitoring of gene expression patterns with a complementary DNA microarray. *Science*, Vol. 270. No. 5235, (Oct 20), pp. 467-470

Schouten J.P., McElgunn C.J., Waaijer R., Zwijnenburg D., Diepvens F. & Pals G. (2002). Relative quantification of 40 nucleic acid sequences by multiplex ligation-dependent probe amplification. *Nucleic Acids Res*, Vol. 30. No. 12, (Jun 15), pp. e57

Schwartz M.K. (1999). Genetic testing and the clinical laboratory improvement amendments of 1988: present and future. *Clin Chem*, Vol. 45. No. 5, (May), pp. 739-745, ISSN 0009-9147

Shaffer L.G., Ledbetter D.H. & Lupski J.R. (2001). Molecular cytogenetics of contiguous gene syndromes: mechanisms and consequences of gene dosage imbalance. In: *Metabolic and Molecular Basis of Inherited Disease*, Scriver C.R., Beaudet A.L., Sly W.S., Valle D., Childs B., Kinzler K.W. & Vogelstein B. (Eds.). pp. 1291-1324, McGraw Hill, New York

Shaffer L.G., Beaudet A.L., Brothman A.R., Hirsch B., Levy B., Martin C.L., Mascarello J.T. & Rao K.W. (2007). Microarray analysis for constitutional cytogenetic abnormalities. *Genet Med*, Vol. 9. No. 9, (Sep), pp. 654-662

Shaffer L.G. & Bejjani B.A. (2009). Using microarray-based molecular cytogenetic methods to identify chromosome abnormalities. *Pediatr Ann*, Vol. 38. No. 8, (Aug), pp. 440-447, ISSN 0090-4481

Udvardi M.K., Czechowski T. & Scheible W.R. (2008). Eleven golden rules of quantitative RT-PCR. *Plant Cell*, Vol. 20. No. 7, (Jul), pp. 1736-1737, ISSN 1040-4651

VanGuilder H.D., Vrana K.E. & Freeman W.M. (2008). Twenty-five years of quantitative PCR for gene expression analysis. *Biotechniques*, Vol. 44. No. 5, (Apr), pp. 619-626, ISSN 0736-6205

Vermeesch J.R., Melotte C., Froyen G., Van Vooren S., Dutta B., Maas N., Vermeulen S., Menten B., Speleman F., De Moor B., Van Hummelen P., Marynen P., Fryns J.P. & Devriendt K. (2005). Molecular karyotyping: array CGH quality criteria for constitutional genetic diagnosis. *J Histochem Cytochem*, Vol. 53. No. 3, (Mar), pp. 413-422, ISSN 0022-1554

Workman C., Jensen L.J., Jarmer H., Berka R., Gautier L., Nielser H.B., Saxild H.H., Nielsen C., Brunak S. & Knudsen S. (2002). A new non-linear normalization method for reducing variability in DNA microarray experiments. *Genome Biol*, Vol. 3. No. 9, (Aug 30), pp. 40-48, ISSN 1465-6914

9

Establishment and Quality Control Criteria for Population Culture Collection - Promising Strategy for Animal Genetic Resource Preservation

Yuehui Ma et al.[1]

Institute of Animal Sciences, Chinese Academy of Agricultural Sciences, Beijing
PR China

1. Introduction

Animal genetic resources top the list of the most fundamental properties for animal husbandry and medical sciences, play an irreplaceable role in human survival and social prosperity, constitute an invaluable substrate for scientific researches, and have an enormous bearing on sustainable development of economy. Biodiversity takes an underlying part in ecological balance. The havoc people wreaked on nature had been aggravating ever since agricultural civilization, accelerating the extinction of animal species and breed incomparable than gradual natural loss, which became more and more apparent upon the advent of industrial age. The statistics of Food and Agriculture Organization (FAO) in 1995 revealed that approximately 15% of the total 738 registered livestock and poultry breeds in Sub-Saharan Africa were on the brink of extinction. The situation has been aggravating ever since. To date, the proportion of livestock breed in danger has increased from 8% to 19%, while that of poultry has risen from 20% to 34%. Among the 1251 registered breeds in Asia, 10% are severely endangered. From 1995 to 1999, livestock breeds about to be extinct grew from 11% to 14%, and the proportion of poultry was 32% to 37%. Owing to economic pressure, some low yield breeds are being subjected to marketing elimination and shrinkage in population, for instance, the production of poultry and swine depends heavily on only a few breeds. The trend is extremely obvious in Eastern Europe, which is further worsened by political unstability. Similarly, sustaining intensification of animal husbandry makes the food production rely more and more on a small number of high yield breeds, thereby exacerbating the animal diversity crisis. In Latin America, the number of endangered breeds accounts for 20% of the whole. As was reported by the FAO in 2000, livestock and poultry throughout the earth are disappearing at the rate of 2 breeds per week. Worse still, 1350 breeds are next to imminent extinction. For all that matter, animal genetic resources are confronted with a progressive narrowing in diversity. Accordingly, it's absolutely exigent to protect and preserve them with effective measures.

[1] Weijun Guan, Xiangchen Li, Xiaohong He, Yabin Pu, Qianjun Zhao, Dapeng Jin, Shen Wu, Taofeng Lu, Xiaohua Su, Chunyu Bai
Institute of Animal Sciences, Chinese Academy of Agricultural Sciences, Beijing, PR China

In the context of biodiversity crisis, no country around the world stands indifferent regarding the preservation of animal genetic resources. The contest for genetic resources, the basis for animal husbandry and sustainable development, is nearly incandescent.

Scientists from all around the world have been endeavouring to preserve and to make use of animal genetic resources, which are now stored in terms of individual animals, semen, embryos, genomic libraries, cDNA libraries, etc. Unfortunately, these alternative methods remain problematic for several reasons: i) endangered species and breeds are incredibly diversified, rendering the costs for preservation of individual animals unaffordable; ii) some core techniques for semen and embryos are still immature; iii) genomic DNA or organ preservation are not applicable in long term because of their finite proliferative capabilities; iv) genomic libraries and cDNA libraries are not the basic unit of cellular activities, moreover, their biological function can only be represented in transgenic techniques. Accordingly, preservation of animal genetic resources in terms of somatic cells is essentially an effective and appealing procedure to protect vulnerable mammalian and avian species, as well as all other kinds of animals. In comparison, somatic cell line, by virtue of its low costs, large capacity, convenient application, proliferative potential and so on, is supposed to be a promising strategy for storage of animal genetic resources.

Consistent with this notion, culture collections of animal materials, mainly identified cell lines, have been established and developed over time. American Type Culture Collection (ATCC), for instance, endeavours to isolate, collect, preserve and supply reliable cell lines, with its all identified type culture applicable for register, preservation, instant use, and even commercial provision. European Collection of Animal Cell Culture (ECACC), jointly run by England and Switzerland, has collected and identified some 1600 cell lines. Established in 1986, Kunming Institute of Zoology, Chinese Academy of Sciences has collected cell stains, tissues and germ cells of numerous precious species of wild life.

The lab of Animal Genetic Resources, Institute of Animal Sciences, Chinese Academy of Agricultural Sciences, by virtue of its solid technical platform, advantages in animal genetic resources, and persistent efforts in the past decade or so, has established Animal Population Culture Collection of China (APCCC), the most massive animal population culture collection in the world, comprising somatic cell lines with different tissue origins, including ear margin, blood, kidney, heart, brain, muscle, stomach and so on, from 88 animal breeds primarily in China, *e.g.* chicken (*Gallus gallus*), duck (*Anseriformes Anatidae*), goose (*Anser cygnoides orientalis*), sandpiper (*Scolopacidae*), cattle (*Bos taurus*), sheep (*Ovis aries*), goat (*Capra hircus*), pig (*Suidae*), ferret-polecat (*Mustela Pulourius Furot*), raccoon dog (*Nyctereutes procyonoides*), horse (*Equus caballus*), mule, red deer (*Cervus elaphus*), sika deer (*Cervus nippon*), fox (*Vulpinae*), wolf (*Canis lupus*), bactrian camel (*Camelus bactrianus*), tiger (*Panthera tigris*), *etc.*, and further endeavours to conserve other animal species and breeds in the world. A well-orchestrated series of standardized technical lines and quality control criteria is steadily ameliorated in this process.

This chapter will introduce the preservation of animal genetic resources in terms of somatic cells and the quality control criteria by detailed experimental description and technical line.

2. Isolation, *in vitro* culture and identification of somatic cell lines

2.1 Sampling and cell culture

Tissue pieces (about 1 cm^3 in size) were sampled from animals and placed into sterile tubes containing Dulbecco's modified Eagle's medium (DMEM, Gibco) (for livestock breeds)/

Modified Eagle's Medium (MEM, Gibco) (for poultry breeds) medium supplemented with ampicillin (100 U/ml) and streptomycin (100 µg/ml). The samples were immediately brought back to the laboratory for further experiments.

The tissues were rinsed and chopped to 1 mm^3 pieces, which were then plated onto the bottom of a tissue culture flask in an incubator at 37°C with 5% CO_2 for 2 h until the tissue pieces spontaneously adhered to the flask surface, and then DMEM/MEM containing 10% fetal bovine serum (FBS, Gibco) was added. Cells were harvested when they reached 80%–90% confluence and were passaged into more flasks at the ratio of 1:2 or 1:3 (Freshney 2000).

After three passages, the cells in logarithmic phase were harvested and resuspended at the concentration of 4 ×10^6/ml in cryogenic media containing 40% DMEM/MEM, 50% FBS and 10% DMSO (Sigma), aliquoted into cryovials, and kept at 4°C for 20-30 min to equilibrize the DMSO. Then they were put into a programmed cryopreservation system with controllable temperature dropping rate, and finally transferred to liquid nitrogen for long-term storage (Jenkins, 1999).

2.2 Trypan Blue exclusion test

Viabilities before cryopreservation and after resuscitation were determined using Trypan blue exclusion test. The cells were plated in 6-well plates at 10^4/well and counted with a hemocytometer (Qi et al., 2007).

2.3 Growth dynamics

According to the method of Gu et al. (Gu et al., 2006) and Ikeda Y et al. (Ikeda Y, 1990), cells at the concentration of 1.5×10^4/ml were plated into 24-well plates. Three wells were counted each day until the plateau phase. Based on the numbers, the mean values of cell density were then calculated and plotted against the culture time. The population doubling time (PDT) was calculated accordingly.

2.4 Microbial detection

- Tests for contamination with bacteria, fungi and yeasts: the cells were cultured in antibiotic free media. Bacterial, fungal and yeast contamination was assessed within 3 days as described by Doyle et al. (1990).
- Test for viruses: the cells were subjected to Hay's hemadsorption protocol using phase-contrast microscopy to detect cytopathogenesis (Hay, 1992).
- Test for mycoplasmas: cells were cultured in antibiotic free media for at least 1 week, and then fixed and stained with Hoechst 33258 (Sigma) according to the method of Masover and Becker (1998) and Freshney (2000). The ELISA Mycoplasma Detection kit (Roche, Lewes, East Sussex, UK) was used to confirm the results of the DNA fluorescent staining.

2.5 Karyotyping and chromosomal indices

The cells were harvested upon 80%–90% confluence. Microslide preparation and chromosome staining were performed as described by Suemori et al. (2006). Fifty to 100 spreads were sampled for counting chromosome numbers of diploid cells. There are three important parameters for chromosomal analysis, i.e. relative length, arm ratio, and centromere index, which were determined according to the protocol of Kawarai et al. (2006).

2.6 Isoenzyme analysis

Isoenzyme profiles of lactic dehydrogenase (LDH) and malic dehydrogenase (MDH) were identified by vertical slab non-continuous polyacrylamide gel electrophoresis (PAGE). The cells were harvested, pelleted and resuspended in protein extraction solution (0.9% Triton X-100, 0.06 mmol/L NaCl:EDTA in volume ratio of 1:15) at the density of 5×10^7 cells/ml. Then the suspension was centrifuged and the supernatant was stored in aliquots at -70°C. Isovolumic 40% (m/v) sucrose and 2.5 ml loading buffer were added to each sample to get loading solution (Zhongxiao and Shuzheng, 1999). Subsequently the electrophoresis was performed at the voltage of 120 V. When the bromophenol blue migrated into the separation gel, the electrophoresis voltage was modified to 220 V. The electrophoresis was terminated when the bromophenol blue migrated to the bottom (0.5 - 1 cm to the margin). Different mobility patterns were differentiated by the relative mobility front (RFs), which was calculated as the ratio of the migration distances of the isozyme bands to that of the bromophenol blue.

2.7 Expression of exogenous genes

According to the method of Tsuchiya et al. (2002), the fluorescent plasmids pEGFP-C1, pEGFP-N3, pEYFP-N1, pDsRed1-N1,pECFP-N1 and pECFP-mito were transfected into the cells with Lipofectamine™ 2000 transfection reagent (Invitrogen Corp., Carlsbad, CA). The plasmid DNA (µg) to Lipofectamine 2000 (µl) ratio was 1:3. After 8 h, the serum-free transfection media were replaced with complete media. To evaluate the transfection efficiency, the cells were observed under a confocal microscope (Nikon TE-2000-E, Japan)at 24 h, 48 h and 72 h after transfection, respectively. The test data were subjected to multiple comparisons to analyze statistical difference. For each sample, images were captured from ten visual fields, and the total and positive cells were counted in each field to determine the transfection efficiency.

3. Results

3.1 Morphological observation

The somatic cells sprouting from tissue explant pieces grew rapidly and migrated from the tissues with a different time and speed according to the species origin. Fibroblasts were initially mingled with epithelial cells, but the fibroblasts, by virtue of their proliferative superiority, would outgrow the epithelial cells gradually after 2-3 passages, and prevail in the population (Fig. 1). Then, purified fibroblast lines were obtained. The cells had fibrous contour with plump cytoplasm, and during growth they were morphologically fibroblast-like with radiating, flame-like or whirlpool migrating patterns. The cells were then subjected to programmed cryopreservation. Trypan blue exclusion test showed non-significant difference ($P>0.05$) in viability upon proper freezing procedures, and resuscitated cells displayed good morphology and proliferative activities.

3.2 Growth dynamics

The growth curves of the somatic cell lines in APCCC before cryopreservation and after cryopreservation displayed a typical "S" shape (Fig. 2) and the PDT was approximately 24 h to 48 h, which varies from species to species or even between subspecies. There is usually a lag time or latency phase of about 24 h to 48 h after plating, corresponding to the adaptation and recovery of the cells from trypsinization, and then the cells proliferate rapidly and enter

Establishment and Quality Control Criteria for Population Culture Collection - Promising Strategy for
Animal Genetic Resource Preservation

169

exponential phase. As the cell density increased, proliferation slows down due to contact inhibition. The cells subsequently enter the plateau phase and begin to degenerate. The growth curves before cryopreservation is generally consistent with that after resuscitation.

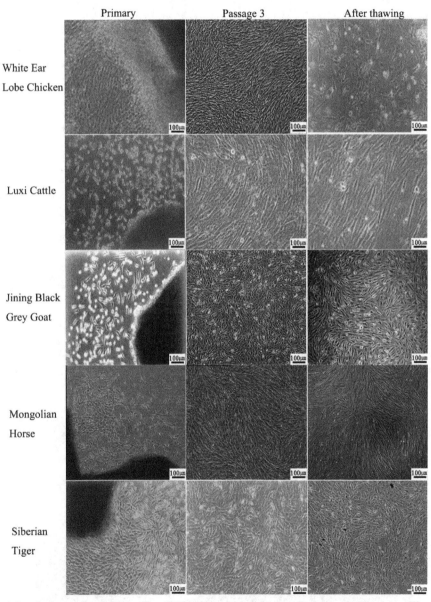

Fig. 1. Morphology of somatic cells of White Ear Lobe chicken, Luxi cattle, Jingning Black Grey goat, Mongolian horse and Siberian tiger in primary culture, before cryopreservation and after resuscitation.

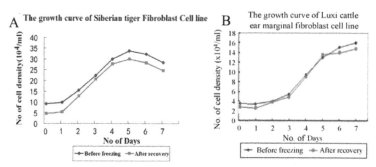

Fig. 2. Growth dynamics. Growth curves of (A) Siberian Tiger fibroblast line, and (B) Luxi Cattle ear marginal fibroblast line before cryopreservation and after resuscitation. A representative growth curve consist of latency phase, exponential phase, plateau phase and decline phase.

3.3 Microbial detection

In a sharp contrast with infections by bacteria, fungi and yeasts, characterized by turbidity, colony or hypha, which can be observed by unaided eyes, the mycoplasma contamination, usually undistinguishable, is only accompanied with slightly slower growth and increased cell fragmentation. As a result, Hoechst 33258 staining or molecular assays are required further. Would there be abundant punctiform and filiform blue fluorescence in the nucleoli, it could be concluded that the cells were contaminated by mycoplasmas (Fig. 3B).

In APCCC, all the somatic cells are subjected to microbial detection prior to cryopreservation to ensure they are free of bacterial, fungal and yeast contamination. No microorganisms were observed in the culture media. No viruses were indicated either by the hemadsorption test. Mycoplasma testing by both the ELISA detection kit and Hoechst 33258 staining (Fig. 3A) was negative. Through microbial detection, the safety and reliability of future commercial and experimental applications of the cell lines are to a large extent ensured.

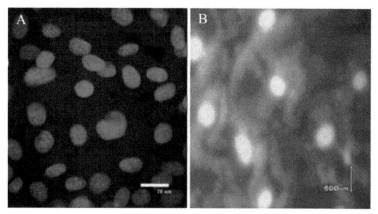

Fig. 3. Hoechst 33258 staining for (A) the detection of mycoplasma in the Siberian tiger fibroblasts; and (B) a positive control of mycoplasma contamination.

Establishment and Quality Control Criteria for Population Culture Collection - Promising Strategy for
Animal Genetic Resource Preservation

171

3.4 Karyotyping and chromosomal indices

Diploid cells of a given species possess a characteristic chromosome number, shape and structure, which remain very stable in normal cells (Fig. 4). Therefore, karyotype analysis is a major method for distinguishing normal cells from mutants. The percentage of diploid cells tends to decrease with increasing passage number. However, the fact that the diploid proportion is normally higher than 90% warrants the hereditary stability.

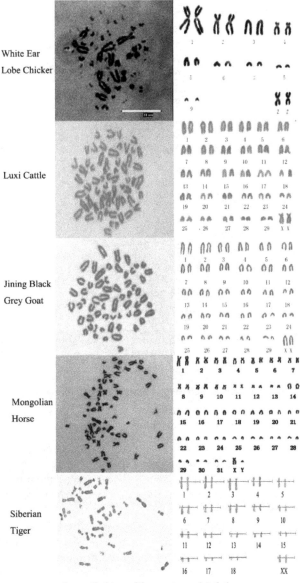

Fig. 4. Chromosome at metaphase (left) and karyotype (right).

Chromosome number	Relative length (%)	Centromere type
1	5.58 ± 0.26	T
2	5.12 ± 0.16	T
3	4.68 ± 0.34	T
4	4.49 ± 0.41	T
5	4.23 ± 012	T
6	4.05 ± 0.45	T
7	3.87 ± 0.38	T
8	3.86 ± 0.57	T
9	3.81 ± 0.04	T
10	3.76 ± 0.22	T
11	3.61 ± 0.11	T
12	3.56 ± 0.19	T
13	3.41 ± 0.33	T
14	3.36 ± 0.20	T
15	3.27 ± 0.41	T
16	3.26 ± 0.32	T
17	3.01 ± 0.09	T
18	2.97 ± 0.19	T
19	2.97 ± 0.06	T
20	2.71 ± 0.31	T
21	2.70 ± 0.24	T
22	2.60 ± 0.12	T
23	2.58 ± 0.27	T
24	2.21 ± 0.19	T
25	2.14 ± 0.22	T
26	2.09 ± 0.53	T
27	2.07 ± 0.10	T
28	1.85 ± 0.35	T
29	1.75 ± 0.32	T
X	4.47 ± 0.11	SM

Note: Relative lenth, 1.0-1.6, Metacentric chromosome (M); 1.7-2.9, Submetacentric chromosome (SM); 3.0-6.0 Subtelocentric chromosome (ST); ≥7.0 Telocentric chromosome (T).

Table 1. Chromosomal parameters of White ear lobe chicken($\female$).

The chromosome number of Luxi cattle was 2n = 60, comprising 58 autosomes and two sex chromosomes, XY or XX. All the autosomes are acrocentric, and only the two sex chromosomes (XY) were submetacentric (Table 1). The chromosome numbers were counted for 100 spreads of passages 1, 3 and 4 respectively, and the frequencies of cells with 2n = 60 were 92.2%, 91.6% and 90.7% accordingly.

The chromosome number of Siberian tiger is 2n=38, consisting of 36 autosomes and two sex chromosomes, XY or XX. The karyotype composition of the Siberian tiger is 12 (M) + 16 (SM) + 4 (ST) + 4 (T), XY (M, M) (Table 2). The chromosome numbers were counted for 100 spreads of passages 1, 3 and 4 respectively, and the frequencies of cells with 2n=38 were 91.6%, 91.2% and 90.2% accordingly.

Chromosome No.	Relative lenth (%)	Centromere type
1	10.13±0.93	SM
2	9.48±1.05	M
3	8.43±0.92	ST
4	6.65±0.85	M
5	6.31±0.81	SM
6	5.85±0.75	ST
7	5.66±0.70	M
8	5.34±0.67	SM
9	5.22±0.71	SM
10	4.47±0.61	SM
11	4.11±0.75	M
12	3.51±0.66	SM
13	3.54±0.44	M
14	3.34±0.69	T
15	3.18±0.67	T
16	2.84±0.25	SM
17	2.43±0.32	SM
18	2.25±0.64	M
X	5.54±0.62	M

Table 2. Chromosomal parameters of Siberian tiger.

3.5 Isoenzyme analysis

Isoenzyme profiles of at least 5 kinds of animals were analysed simultaneously. Each kind of animal has its specific bands. The LDH bands obtained from Siberian tiger, for instance, were compared with those of other species or breeds, and five breed-specific isoenzyme bands (LDH-1, -2, -3, -4, -5) were observed (Fig. 5A). Enzymatic activities were in the order of LDH-3, LDH-2, LDH-4, LDH-5, LDH-1. LDH-2, LDH-3 and LDH-4 were dominant, while LDH-1 and LDH-5 were scarcely observable. In the MDH patterns of Siberian tiger and other breeds, two MDH bands (s-MDH, m-MDH) were observed (Fig. 5B), with the m-MDH band near the cathode and the s-MDH band (comprise two subbands but hardly identified) near the anode (Fig. 5B). Similar activity was seen from both m-MDH and s-MDH. There were significant differences in the isoenzyme patterns of LDH and MDH between the Siberian tiger fibroblasts and other cell lines in APCCC. These animals have their distinctive bands with different relative mobility. These results showed that there was no cross-contamination between different cell lines.

3.6 Expression of exogenous genes

Six fluorescent protein genes with stable structures, high expression levels and species-independent efficiency (Baird et al., 2000) have been used as marker genes to observe the expression, distribution and function of target proteins in live cells and organisms (Heim et al., 1995; Genyang et al., 2003). At APCCC, the 6 kinds of fluorescent genes were introduced into the preserved cells to evaluate the expressibility of exogenous genes. Positive cells were usually the most abundant and with the strongest fluorescence at 24 h-48 h after transfection. While the transfection efficiency decreased, strong expression levels were observed after a week, indicating that the exogenous genes can be replicated, transcribed,

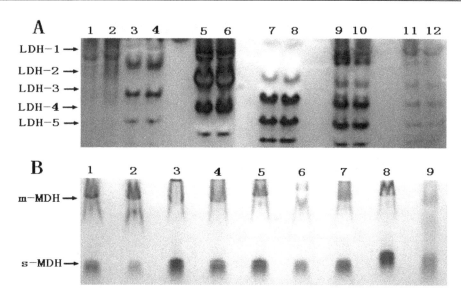

Fig. 5. LDH zymotype and MDH zymotype of several cell lines. A, SDS–PAGE electrophoresis of LDHs, from up to down, there were LDHs-1, 2, 3, 4 and 5. Panel A: 1,2 Simmental cattle, 3,4 Zhiwei goat, 5,6 Jining black goat, 7,8 Mongolian horse, 9,10 Bengal tiger, and 11,12 Siberian tiger; Panel B, MDHs from up to down were mMDH and sMDH. 1 Siberian tiger, 2 Bengal tiger, 3 Large white pig, 4 Songliao Black pig, 5 Jining Black goat, 6 Mongolian sheep, 7 Saf sheep, 8 Simmental cattle.

translated and modified within the cells. The transfected cells were not significantly less viable than the control cells ($P>0.05$), showing that the expression of fluorescent proteins had no obvious effect on the growth and proliferation of the transfected cells.

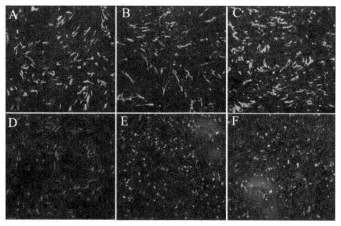

Fig. 6. Comparative figures of six fluorescent proteins in White Ear Lobe chicken fibroblasts at 24 h after transfection (×10). A, B, C, D, E and F were the transfection results of pEGFP-C1, pEGFP-N3, pEYFP-N1, pDsRed1-N1, pECFP-N1 and pECFP-mito, respectively.

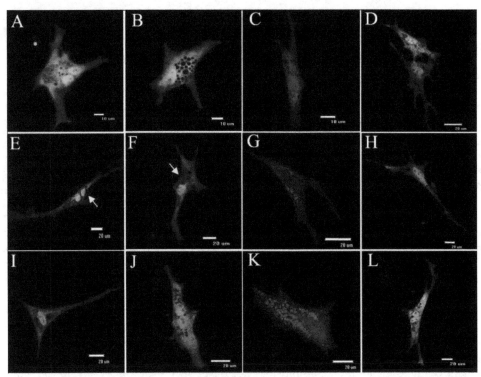

Fig. 7. The expression and distribution of pEGFP-C1, pEGFP-N3, pDsRed1-N1 and pEYFP-N1 in White Ear Lobe chicken fibroblasts (×40). A, B, C and D are the expression of pEGFP-C1, pEGFP-N3, pDsRed1-N1 and pEYFP-N1 at 24 h; E, F, G and H at 48 h; and I, J, K and L at 72 h after transfection.

4. Conclusion

Animal resources, a fundamental respect of agriculture and industry in close correlation with production and social stability, supply human beings with meat, eggs, milk, furs, medicinal materials, products for athletic and ornamental purposes, etc. In most developed countries, scalization of animal husbandry has restricted animal feeding to high yield breeds or crossbreeds with an intensified operating system, greatly compromising the diversity of local animal breeds. In the meanwhile, despite the abundance of animal genetic resources in developing countries, the lack of efficient preservation strategies and blind introduction of exotic breeds for hybridization also has reduced the animal variety. Emerging evidence has revealed that owing to the interference from human activities, species extinction has sped up for about 1000 fold, 100 million times faster than speciation, or in other words, 1 species per day. The total 7176 livestock and poultry breeds throughout the earth are disappearing at the rate of 2 per week, and 690 are on the edge of extinction.

Species extinction signifies a perpetual loss of the precious hereditary information, and will be an irreparable defeat of world genetic resources and biological theoretical repositories. Haven't the genetic resources been preserved in any forms before their extinction, not only the genetic resources will be lost evermore, but also it becomes impossible to investigate the unknown cell and molecular mechanisms, let alone to regenerate corresponding species via cloning technique. Therefore, it is exigent to employ practical measures to conserve endangered animal species.

Therefore, the APCCC has as yet preserved, for each cell line, 45-1250 cryovials of somatic cells from 30-212 individuals using primary explantation, serial passage and programmed cryopreservation. Each vial contains approximately 1.5×10^6 cells. The cells are cryopreserved within 3 passages, and are subjected to evaluation in terms of morphology, growth kinetics, viability, microbial detection, karyogram and isoenzyme analyses according to quality control standards of ATCC. The purified fibroblasts are fusiform, displaying flame-like or swirl-like patterns. The growth curves are sigmoidal with characteristic PDTs. Trypan blue exclusion test suggests that programmed preservation exerts a non-significant effect ($P>0.05$) on cell viability compared with that before freezing. Tests for bacteria, fungi, viruses and mycoplasmas are unanimously negative. Karyograms of peripheral blood lymphocytes and the in vitro cultured cells are photographed, according to which the mode of chromosome numbers are determined as that of the diploid cells, and indices including relative length, arm ratio and centromeric index and kinetochore type are calculated or dertermined. Comparison between peripheral blood lymphocytes and the in vitro cultured cells in respects of chromosome number and non-banding karyotype reveals no perceptible differences, manifesting the genetic stability of the cell lines established. Isoenzyme patterns of LDH and MDH are detected using vertical slab non-continuous PAGE assay, the breed specific bands of which rule out cross-contamination amongst the cell lines, and in the meanwhile further evince the hereditary stability.

Aforementioned results indicated that the APCCC conforms to all the ATCC criteria for somatic cell lines. In addition, plasmids of pEGFP-N3, pEGFP-C1, pECFP-N1, pECFP-mito, pDsRed1-N1, and pEYFP-N1 encoding the corresponding fluorescent proteins are transfected into the cells using lipofectin mediated protocol to study the expression of exogenous genes. By observation or detection of spatiotemporal expression of the fluorescent proteins, proliferation and growth of positive cells, apoptotic rate and viability, the ability and characteristics to accommodate exogenous genes are initiatively adopted as a constitutional index for cell line quality control.

The establishment of the APCCC is technically and theoretically conducive to preserve genetic resources of animals at somatic cell level, and definitively has a profound and long-lasting influence on biological and biomedical research in the future. The quality control standards it's been adopting will definitely provide insights for future development of culture collections.

5. Abbreviations

APCCC - Animal Population Culture Collection of China
ATCC - American Type Culture Collection
DMEM - Dulbecco's modified Eagle's medium
ECACC - European Collection of Animal Cell Culture
FAO - Food and Agriculture Organization
LDH - lactic dehydrogenase
MDH.- malic dehydrogenase
MEM - modified Eagle's medium

Establishment and Quality Control Criteria for Population Culture Collection - Promising Strategy for
Animal Genetic Resource Preservation

177

PAGE - polyacrylamide gel electrophoresis
PDT - population doubling time

6. References

Baird GS, Zacharias DA, & Tsien RY. (2000). Biochemistry, mutagenesis, and oligenerization of DsRed, a red fluorescent protein from coral. Proc Natl Acad Sci. Vol. 97, No. 22, pp. (11984-9). ISSN 0027-8424

Doyle A., Hay R., & Kirsop B.E. (1990). Animal Cells (Living Resources for Biotechnology). Cambridge University Press, ISBN 0521352231, Cambridge, UK.

Freshney R.I. (2000). Culture of animal cells: a manual of basic technique(4th ed). Wiley-Liss,Inc., ISBN 0471348899, New York.

Genyang Cheng, Xiangmei Cheng, & Xueyuan Bai. (2003). The gene construction and location in tubular epithelial cells of fused by green fluorescence protein and human kidney gene NaDC3. J Cell Biol. Vol. 25, No. 3, pp.(170-3). ISSN 1540-8140.

Gu Y.P., Li H.Z., & Mik J. (2006). Phenotypic characterization of telomeraseimmortalized primary non-malignant and malignant tumor-derived human prostate epithelial cell lines. Experimental Cell Research. Vol. 312, No. 6. Pp. (841–843). ISSN 0014-4827.

Hay R.I., (1992). Cell line preservation and characterization. In: Animal Cell Culture: A Practical Approach (2nd ed). Freshney R.I. pp. (104–135), Oxford University Press,. ISBN 0199637962,Oxford.

Heim R, Cubitt AB, & Tsien RY. (1995). Improved green fluorescence. Natur. Vol. 373, No. 6516, pp. (663-4).ISSN 1476-4687.

Jenkins N. (1999). Animal cell biotechnology methods and protocols. Humana Press, ISBN 0896035476, New Jersey.

Kawarai S., Hashizaki K., & Kitao S. (2006). Establishment and characterization of primary canine hepatocellular carcinoma cell lines producing alpha-fetoprotein. Vet. Immunol. Immunopathol. Vol. 113, No. 1-2, pp. (30–36), ISSN 0165-2427.

Ikeda Y, Ezaki M, Hayashi I, Yasuda D, Nakayama K, & Kono A. (1990). Establishment and characterization of human pancreatic adenocarcinoma cell line in tissue culture and the nude mouse. Jpn J Cancer Res. Vol. 81, No. 10, pp. (987-93). ISSN 0910-5050.

Masover G.K., Becker F.A., (1998). Detection of mycoplasmas in cell cultures by cultural methods. In: Methods in Molecular Biology, Miles R.J., Nicholas R.A.J., et al. pp. (207–215, 217–226). Humana Press Inc., Totawa NJ.

Qi Yitao, Tu Yiding, Yang Di, Chen Qian, Xiao Jun, & Chen Yiqiang. (2007). Cyclin A but not cyclin D1 is essential in c-myc-modulated cell cycle progression. J Cell Physiol. Vol. 210, No. 1, pp. (63-71). ISSN 1097-4652.

Suemori H., Yasuchika K., Hasegawa K., Fujioka T., Tsuneyoshi N., & Nakatsuji N. (2006). Efficient establishment of human embryonic stem cell lines and long-term maintenance with stable karyotype by enzymatic bulk passage. Biochem. Biophys. Res. Commun. Vol. 345, No. 3, pp. (926–932), ISSN 0006-291X.

Tsuchiya R., Yoshiki F., Kudo Y., & Morita M. (2002). Cell type-selective expression of green fluorescent protein and the calcium indicating protein, yellow cameleon, in rat cortical primary cultures. Brain Res. Vol. 956, No. 2, pp. (221–229), ISSN . 0006-8993.

Zhongxiao He, & Shuzheng Zhang. (1999). Electrophoresis. Scientific Press; Beijing.

Part 4

Planning for Quality Control

10

QA/QC in Transport Infrastructures: Issues and Perspectives

Filippo G. Praticò
University Mediterranea at Reggio Calabria
Italy

1. Introduction

Transport infrastructures (road and highway networks, railways, terminal facilities, airports, mass transit systems, bicycle paths and pedestrian walkways, etc.) have significant impacts on the development of regional and national economies. However, the effectiveness of these impacts over the time has to be established based on the actual quality of all the pertaining components of the infrastructure asset (pavement, safety barriers, signals, illumination, embankment, drainage, etc.). Quality can be interpreted as the degree of excellence of a product or service, or as the degree to which a product or service satisfies the needs of a specific customer or, finally, as the degree to which a product or service conforms with a given requirement. In more detail, **quality assurance** (QA) refers to all those planned and systematic actions necessary to provide confidence that a product or facility will perform satisfactorily in service. At the same time, **quality control** (QC), also called process control, relates to those QA actions and considerations necessary to assess and adjust production and construction processes so as to control the level of quality being produced in the end product (Fig.1). Note that **QA (which includes QC) is an infrastructure (e.g. highway) agency responsibility** and involves all the process (planning, design, plan and specifications, construction, etc.), while **QC is a producer–contractor** responsibility which mainly affects construction. Furthermore, QC is not simply QA in construction, due to the fact that both independent assurance and acceptance procedures refer to QA in construction but they are NOT a part of QC. The entire QA/QC process includes: i) setting up the initial data collection or experimentation to determine typical parameters of current construction; ii) designing the acceptance plan itself, including selecting quality characteristics (and corresponding specification limits), statistical quality measure (and corresponding quality levels), buyer's and seller's risks, lot size, number of samples (sample size), specification and/or acceptance limits, and payment–adjustment provisions. As is well known (Burati et al, 2003), traditionally, highway specifications spelled out in detail the work that was to be done by the contractor under the form of **materials and methods specifications** (also called method specifications, recipe specifications, or prescriptive specifications). In this case, specifications direct the contractor to use specified materials in definite proportions and specific types of equipment and methods to place the material. On the contrary, **end result specifications** require the contractor to take the entire responsibility for supplying a product or an item of construction.

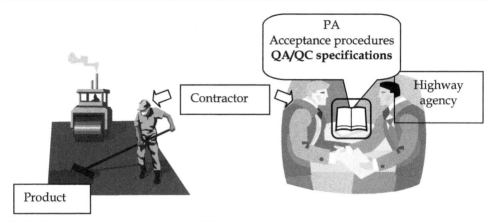

Fig. 1. Contractor vs. agency relationship.

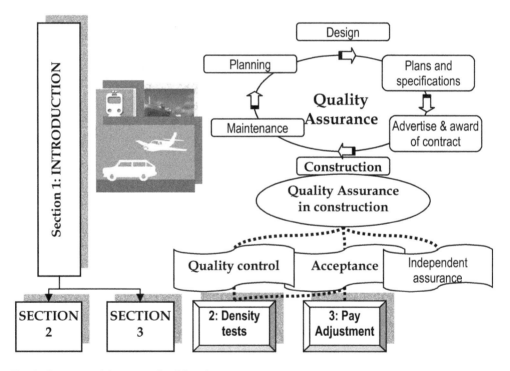

Fig. 2. Conceptual framework of the chapter.

The highway agency's responsibility is to either accept or reject the final product or to apply a price adjustment (PA) commensurate with the degree of compliance with the specifications. In practice, current specifications are neither solely "materials and methods" nor "end result." (Burati et al., 2003). **Quality assurance specifications** (a.k.a. QA/QC specifications or QC/QA specifications) are a combination of end result specifications and materials and methods specifications. **QA specification** consists of two separate functions, i.e. **quality control** or process control, and **acceptance**. As for the development of QC procedures and requirements, the QC procedures and requirements are made up of two parts: the QC requirements (tests to be performed, minimum frequency, qualified laboratories) and the quality characteristics to be measured. For HMAC (hot mix asphalt concrete, or HMA), typical quality characteristics that may be tested for QC include aggregate quality, density (see next section), gradation of critical sieve sizes, plant and discharge temperatures, degree of aggregate coating, moisture content of fine aggregate and/or of finished mix. For PCC (Portland cement concrete) , typical quality characteristics that are tested for QC include aggregate quality, gradation of critical sieve sizes, air content, water–cement ratio, mix temperature, slump. Note that three different subjects can carry out the acceptance testing: contractor, agency, third part. The agency: i) may decide **to do (itself)** the acceptance testing; ii) may assign the testing to the **contractor**; iii) may have a **combination of agency and contractor acceptance testing**; iv) or may require a **third part** to do the testing. In Italy, acceptance testing is usually carried out by a third part, following "general" and "special" Contract specifications (Capitolato special CIRS, Capitolato speciale prestazionale ANAS, etc.). Figure 2 illustrates the conceptual framework of the chapter.

2. Density

By referring to density (more technically: bulk specific gravity and air voids content), this is a very crucial factor for QC/QA procedures used to evaluate contract specifications. Indeed, the life cycle of hot mix asphalt (HMA) depends on the material density. Bulk specific gravity, Gmb, measures the specific gravity of a compacted hot mix asphalt sample (core or in-lab compacted). Life cycle costs, contract requirements, and QC/QA procedures are all modeled as functions of the effective Gmb properties resulting from suitable design and construction techniques and by application of appropriate boundary conditions. A variety of methods for determining (in laboratory or on site) Gmb are available (dimensional, AASHTO T 269-EN 12697-6:2003; parafilm, ASTM D 1188; vacuum sealing principle, ASTM D 6752; paraffin coated, BU N40-1973, AASHTO T 275-A, EN 12697-6:2003; saturated Surface Dry, AASHTO T 166, ASTM D 2726, EN 12697-6:2003; non-nuclear portable device, ASTM D 7113, AASHTO TP68). In the vacuum sealing method (VSD), specimen volume is determined by a vacuum chamber that shrink-wraps the specimen in a high quality plastic bag. Surface texture effects can be appreciable, and are accounted for in by the shrink-wrap process. Archimedes' principle is then applied. The dimensional method uses height, diameter, and width measurements to estimate the volume. Surface irregularities (i.e., the rough surface texture of a typical specimen) can introduce inaccuracies, because, in practice, an "osculatory" volume is computed. The parameter P is a density estimate derived from measurements of a non-nuclear portable device, collected at the tested surface (Williams, 2008; Megali et al., 2010; Rao et al., 2007; Kvasnak et al., 2007; Romero, 2002; Sargand et al., 2005; TransTech, 2003; Gamache, 2004, 2005; Praticò et al, 2009; Praticò et al, 2009a; Praticò and Moro, 2011; Alvarez et al, 2010). In non-nuclear portable device measurements, an electrical current is transmitted from a transmitter through the asphalt concrete pavement at a given location and is detected by the receiver. Note that the current cannot flow through the isolation ring. The impedance (ohm) is measured, and the dielectric constant (dimensionless) is derived. The dielectric constant of the HMA is used to estimate the density. The dielectric constant depends on the HMA composition in that it is derived from the dielectric constants of air (~1) and bitumen &

aggregates (5–6). Water and moisture, if present, strongly influence the dielectric constant estimate because the dielectric constant of water approximately 80. Densities measured in the laboratory (dimensional, parafilm, vacuum sealing principle, paraffin coated, saturated surface dry), when contract specifications are well-defined, are quite reliable and accurate but are obtained very slowly. On-site measurements (non-nuclear portable device) are made using non-nuclear portable devices and result often biased and unreliable. Researchers have attempted to find correlations between the results obtained using different procedures (Brown et al., 2004; Cooley et al., 2002; Crouch et al., 2003; Mohammad et al., 2005; Montepara & Virgili, 1996; Spellerberg & Savage, 2004; Williams et al., 2005; Megali et al., 2009). Studies have been carried out in order to propose a theoretical framework for interpreting in-lab and on-site measurements and in the aim of proposing strategies for using non-nuclear portable devices in QC/QA (Megali et al., 2010). A decrease in porosity (or effective porosity) yielded in-lab specific gravities that converged. In contrast, due to the high dielectric constant of water the divergence of in-site measures was observed. The ranking among procedures was the following: GmbSSD > GmbFIN > GmbCOR > GmbFILM > GmbDIM. Furthermore the ranking did not depend on core diameter (Praticò and Moro, 2011b). From a practical standpoint, two strategies were proposed for estimating the density of an asphalt sample: i) consideration of both W (water content) and P (density measured through non-nuclear portable devices) in an estimate for the effective porosity, which is the most crucial parameter for quality control evaluation; ii) consideration of both W and P in an estimate of density.

In (Praticò and Moro, 2011), two equations were proposed for practical applications:

$$G_{mbCOR} = aP + bW + c \tag{1}$$

$$n_{eff} = dP + eW + f \tag{2}$$

where a = 0.914, b = -0.007, c = 0.303, d = -56.673, e = 0.420, f = 128.698 (coefficients are case-specific).

3. Acceptance procedures

This section deals with acceptance procedures. Premises (§3.1) illustrate how acceptance procedures can be carried out, the main indicators which are used and what are the main issues. Afterwards, section 3.2 (model) describes a model for the derivation of the price adjustment (PA) based on life cycle cost analysis. Due to the need of considering surface properties, section 3.3 addresses life expectancy of surface properties (skid resistance, texture, surface wear, drainability, acoustic performance). Bulk properties still remain the key-factor in determining the expected life of the as-constructed pavement. To this end, section 3.4 deals with several tools for the determination of the expected life of a pavement through empirical to rational models. Section 3.5 presents an example of application and provides other references. Main findings are summarized in section 3.6.

3.1 Premises and symbols

In the aim of providing a useful indicator for acceptance procedures, from each quality characteristic the corresponding value of quality measure is derived (Burati et al., 2003; Leahy et al., 2009). Quality measure is any one of several means that have been established to quantify quality. Some examples of quality measures are the mean, the standard deviation, the percent defective, the percent within limits, the average absolute deviation, and the quality index. Note that it is required to determine specification limits (USL, upper specification limit, and LSL, lower specification limit), and to decide on AQL (Acceptable quality level) and RQL (Rejectable quality level). USL and LSL are the limiting values placed on a quality characteristic, while AQL and RQL refer to the quality measure and are respectively the minimum level of actual quality at which the material or construction can be considered fully acceptable (for that quality characteristic) and that maximum level of actual quality at which the material or construction can be considered unacceptable (rejectable). Specification limits and quality levels are basic inputs to decide pay relationships. Performance-related pay, incentive/disincentive, minimum pay provisions, remove/replace provisions, retest provisions are examples of pay relationships. At the present several classes of models for acceptance procedures depending on input parameters and underlying principle can be listed (Praticò, 2007; Praticò et al, 2008; Praticò et al, 2010a; 2010b; Praticò et al, 2011a): 1) IRI-type models (where IRI stands for International Roughness Index) and average-type models; 2) PD-type models (where PD stands for Percent Defective); 3) LCCA based models (where LCCA stands for Life Cycle Cost Analysis). IRI-type models are based on roughness indicators, as synthetically expressive of the quality of the work and of the related costs (for example the Profile Index (PI), the International Roughness Index (IRI), etc.. In this case the Pay Adjustment (PA) is often empirically determined on the basis of the Pay Factor (PF). PF (and the payment to the contractor) decreases as IRI increases. In the average-based models (see for example typical Italian contracts such as CIRS and ANAS 2008), the pay adjustment (or/and its specific value, $PA/C=PA^*$) usually depends on the difference (DI), for the given j-th quality characteristic, between the average (x_{AC}) and the design value (x_{AD}). Sometimes the percent difference is used (%, $PDI=(x_{AC}-x_{AD})/ x_{AD}$)or its difference from a threshold. It results $PA=abs(DI) \cdot k$ (where abs(DI) indicates the absolute value of DI, while k usually ranges from 0.3 to 3 and the overall PA is the sum of the PAs of the single quality characteristics. For example, for a SFC (side force coefficient) of 36 instead of 40, it results PDI=10% and PA=10·0.5=5%. Note that these models are cumulative and only penalties are involved (no bonuses). Thickness, HMA moduli, pavement bearing, surface properties (such as SFC and sand height, SH) are usually the main quality characteristics which are considered. In PD-type models asphalt content, grading, mix in-place density, air voids, Marshall resistance, and/or thickness are often used (Deacon et al., 1997; Epps et al., 1999; Khaled, 2003; Burati, 2005; Hughes et al., 2005). In this case the PA is often (but not always) computed by taking into account the percentage within the limits (PWL), with particular probabilistic hypotheses in order to calibrate agency and contractor risks; for the i-th variable, for a given layer, from the PWLi a percent defective (PDi) and then a pay factor (PFi) are computed; by combining all the PFi a combined pay factor (CPF) is often derived. Given that, by the means of CPF and cost, the Pay Adjustment is computed. LCCA based models are models in which the concept of life cycle cost analysis (as a methodology useful to compare the total user and agency costs of competing project implementation alternatives) is explicit. Therefore, pay adjustment depends on how long the pavement will last (Weed, 2001; Weed & Tabrizi, 2005; Whiteley et al., 2005). The Pay Factor may be often referred (more or less implicitly) to an expected life. For this reason, the boundary between these different models may be difficult to define. Though many algorithms do exist for assessing pay adjustment, many classes of issues still call for research. Furthermore, note that the real OC curves (Operating Characteristic Curves) and risks will depend on sample size (number of test results used to judge the quality of a lot), lot size and sublot size (sublots are needed in order to ensure that the specimens for the sample are obtained from throughout the lot, and are not concentrated in one portion or section of the lot). The operating characteristic (OC) curve is a graphic representation of an acceptance plan that shows the relationship between the actual quality of a lot and either the probability of its acceptance or the probability of its acceptance at various payment levels (for acceptance plans that include pay adjustment provisions). OCs aid in the selection of plans that are effective in reducing risks, because they provide buyer's and seller's risk.

In the abovementioned processes (state-of-the art in the field) the following critical issues can be listed (Praticò, 2008). Problem 1. As is well known (Di Benedetto et al., 1996; Domenichini et al., 1999), all the properties and characteristics influence the real and perceived economic value of the as-constructed pavement at a given time. So, when one characteristic fails, this constitutes a quality assurance problem for the state agency (Burati et al., 2003; Muench & Mahoney, 2001), and an acceptance plan, with particular acceptance procedures, is needed, in order to estimate a pay adjustment, PA. The use of **road surface condition** measurements for the acceptance of roadwork is becoming more and more relevant (Boscaino and Praticò, 2001; Boscaino et al, 2005) and calls for a synergistic approach. Problem 2. Can a friction course be treated as the remaining layers in estimating pay adjustment? In other terms, how can pay adjustment be estimated when both surface and mechanical defects are involved? Problem 3. Is it possible to apply life cycle cost analysis when **both surface and mechanical performance are involved**? Problem 4. Attention is often entirely focused on the quality of single asphalt layers, without any systemic consideration of the quality of the whole **multilayer** during the life cycle. However, it must be observed that the influence of the bottom layers on the performance of a multilayer can strongly modify both failure typology and pavement expected life. Problem 5. **Percentage of defects or average values**? Is there a connection between models based on percent defective and model/procedures based on the consideration of average values? Is there the potential for a synergetic consideration of both position and dispersion? Problem 6. Is there a **relationship between the position of the mean respect to the limits and the value of the percent of defects**? Probably yes, but it depends on the single type of set data (standard deviation, asymmetry, kurtosis, or in the simplest case on standard deviation). Many uncertainties in this field call for further research (Uddin et al, 2011). Problem 7. Given that performance are usually related to averages and not to percent defective, is it possible to relate **pavement performance and PDs** (percentage of defects)? Problem 8. The logic of percent defective is usually linked to pay factors and a **composite pay factor** must be considered. A possibility is to have a correlation between the different factors and the expected life of the pavement. But how much logical and exhaustive can be such procedures? Problem 9. Is it possible a synergetic consideration of **defects and delays** in pay adjustment models? Problem 10. A drawback of the method of percent defects is the impossibility for taking into account the extended service life due to the surplus in some of the quality indicators (for example thickness), even if other defects are detected. In other terms, **PDs are defined positive**. Is there any possibility to correct this shortcoming within the framework of the PD-models? Problem 11. Another problem does occur when one tries to apply the concept of percent defective to Open Graded Friction Courses or Porous European Mixes. In fact, in these cases, the quality indicators used in PD evaluation (thickness and air voids) seem not to provide a logical estimate of expected life (included the surface performance). This fact is due also to the uncertainties in the field of the

$\tau, \tau\tau$	τ refers to each periodical effect on friction (F); $\tau\tau=\max\tau$
α, β, χ	real numbers, coefficients
ϕ_i	i-th phase of the i-th periodical effect on friction (F)
a, a_i, a_{1B}	structural layer coefficients; i refers to i-th layer and B refers to the Bearing component
B	"bearing" component of the pavement
BPN	British Pendulum Number
$C_B, C_S C_{DP} C_i, C_{1B} \% C_B$	Present Costs, referred to "B" characteristics, to "S" ones, to DP, to the i-th layer, to the 1st layer of the "B" component, to "B" characteristics as a percent of C_{DP}, respectively.
CBR	California Bearing Ratio
CP	as-Constructed Pavement, actual pavement constructed by the contractor
CPF	Combined Pay Factor
CT	equivalent Cumulative Traffic
D	Design life of the as-Designed Pavement; also called initial design life, it is the amount of time for which the chosen pavement design is expected to carry the traffic loads without the application of a global rehabilitation.
DFC	Dense-graded Friction Course
DP	as-Designed Pavement; desired pavement, as defined by the agency (buyer)
$\delta_{AC,I}, \delta_{AD,j}$	Parameters which take into account for successive resurfacings.
E, EB, ES	Expected life of the CP, general, for only B component, for only S component, respectively
ESALs	Equivalent single Axes Loads
EXPLIF	Expected life
f, f_3	real numbers, ratio between C_2 and C_1 (f), and ratio between C_3 and C_1 (f3), respectively
F_1, F_2, F_3, F	Coefficients F_i in the formula of the friction F
fre_i	i-th frequency (formula for skid resistance dependence on time)
$G_0, .., G_3$	Real coefficients in the recipe of the expected life (Burati et al, 2003)
h	real number, ratio between t_{1B} and t_2
I	Indicator, for example percolation speed or drainability
INF, INT	Inflation rate and interest rate, respectively
IRI	International Roughness Index
M1B, M1	Moduli, respectively, of the first layer of DP– B component and of the first layer of the DP
MS	Marshall Stability
N, n	Number of layers (above the subgrade) total (N) and to resurface/construct (n)
NDT	Non Destructive Test
O, O_{DP}, O_{CP}	Expected life of successive resurfacing/reconstruction, general, of DP, of CP, respectively
OGFC	Open Graded Friction Course
P	Pavement
PA, PA$_B$, PA$_S$, PA%, PA%(1L),..	Pay Adjustments; PA is the total one; B and D mean referred to B or S, respectively; %:expressed in percentage, i.e. referred to CDP; (1L): referred to one layer (1L) or more (2L, etc.)
PD_V, PD_T	Percent Defective (PD) referred to air voids and to thickness respectively.
PEM	Porous European Mixes
PF	Pay Factor
PI	Profile Index (a roughness indicator)
PM	Preventive Maintenance

PWL	Percentage Within Limits
PWL	Percentage Within Limits
REH	Rehabilitation
S	supplementary component of the pavement. It hasn't bearing properties.
SFC	Side Force Coefficient
t_i t_{1B}	Thickness of the i-th layer of DP and of the first layer of the bearing component of DP
T_j	expected life of the j-th supplementary characteristic
UCS	Unconfined Compressive Strength – 7 day break
Y, Y*	Years (real number) and time in years to reach a quasi-constant friction, respectively

Table 1. List of symbols and acronyms.

correlations among air voids, moduli and expected life. Problem 12. Can base and subgrade be part of the considered pavement system in such algorithms? Problem 13. The density and volumetrics of as-built pavements is a vital part of QC/QA procedures.

Expected life, infrastructure management and pay adjustment strongly depend on air voids content, especially when bituminous mixes are involved. Despite this measurement process is affected by several classes of uncertainties and many issues still call for further research: influence of core diameter, reliability of non-destructive testing, etc. Table 1 lists the main symbols used in this section.

3.2 Model

This section deals with model development. Equations 3-13, Fig.3, Table 1 summarize the derivation of the algorithms. Note that the algorithm here presented overcomes and refines the previous formula as stated in (Praticò, 2007) and was successively updated (Praticò et al, 2010a; Praticò et al, 2010b; Praticò et al, 2011). In summary, the model allows to estimate the pay adjustment on the basis of how long the pavement (considered in all its "qualities") will perform adequately. In order to introduce the model, let DP be the as-designed pavement and CP be the as-constructed pavement (symbols are listed in table 1). The main inputs of the model are costs and expected lives, while the output is the Pay Adjustment, PA (negative if a penalty is provided). It is well known that the friction course has supplementary characteristics (friction, fire resistance, etc., Praticò et al, 2010c). Therefore, every pavement P (both DP and CP, for example) can be divided into two main "components": Bearing characteristics, B, and Supplementary characteristics, S. By comparing, separately, the Bearing characteristics (B, addressing substantially moduli, Poisson coefficients and thicknesses) of the as-Designed Pavement (DP) and of the as-Constructed Pavement (CP), the Pay Adjustment PA_B is estimated (where the subscript B means that PA is referred to the bearing characteristics). Similarly, by comparing the Supplementary characteristics (S) of as-designed (DP) and of the as-constructed pavement (CP), the pay Adjustment PA_S is estimated:

$$P = B + S \tag{3}$$

$$PA = PA_B + PA_S \tag{4}$$

where PA_S refers to S, PA_B to B, and PA to all the non-conformities. To estimate PA_S and PA_B, it is necessary to analyze the costs of the pavement during its life. Let N be the total layers of the as-Designed Pavement (above the subgrade) and let n be the layers to be resurfaced (or reconstructed, if base layers are involved): thus $n \leq N$. Each of the n layers has a contract cost (i.e. present cost of the as-Designed Pavement, in €/m²) equal to C_i (i=1, 2, .. n, from the surface towards the subgrade), then the relative cost C_{DP} of the as-Designed Pavement can be expressed as follows:

$$C_{DP}=C_1+C_2+...+C_n. \tag{5}$$

$$C_{DP}=C_S+C_B \tag{6}$$

where C_S and C_B are respectively the cost of the S component (supplementary) and of the B component of the as-designed pavement. The first layer of B will have a present cost C_{1B} (where C stands for cost, 1 for 1st layer, B for bearing component of the pavement) and a thickness t_{1B} (in order to permit to B to last for D years). Thus:

$$C_B= C_{1B}+C_2+...+C_n. \tag{7}$$

$$C_S=C_{DP}-C_B=C_1-C_{1B} \tag{8}$$

where C_{1B}, referred to the first layer of the B component (which is intrinsically designed only to have mechanical properties), is generally lower than C_1. Let INT and INF be the long-term annual interest rate and inflation rate respectively, given in decimal form (typically 0.08 and 0.04). The Rate R is so defined:

$$R=(1+INF)\cdot(1+INT)^{-1}. \tag{9}$$

For the as-Designed Pavement, let Design life (D) be the expected life in years of the B component and D_S of the S component. For the as-Constructed Pavement, let E_B be the expected life of the B component and let E_S be the expected life of the S component. In practice, E_S can be interpreted as the minimum expected life (years) for supplementary characteristics (where j represents a given supplementary characteristic):

$$ES=min\ [ESi],\ i=1,\ 2,\ ...,\ k \tag{10}$$

where ES_i is the expected life of the i-th supplementary characteristic. Note that, in order to consider the right number of resurfacing processes the parameters δ ($δ_{AC}$ and $δ_{AD}$) can be introduced, where, for example, a=1, m=1, n=20000, τ=0.05, t=EB-ES-DS:

$$δ=a\cdot(1+m\,e^{-t/τ})\cdot(1+n\,e^{-t/τ})^{-1}. \tag{11}$$

In practice, for t=EB-ES-DS=0 or negative, δ approaches 0, while for EB-ES-DS>1 it approaches 1. Note that in a first analysis of the problem δ can be neglected. Let us introduce the concept of expected life (years) of successive resurfacing or reconstruction (typically 10 years). For the as-Designed Pavement (DP) let ODP be this "successive" expected life, both for B (bearing) and S (supplementary). For the as-Constructed Pavement (CP) let OCP be this "successive" expected life, both for B (bearing) and S (supplementary). The reason for separating the concept of O into two different concepts ODP and OCP is that the actual expected life of a resurfacing / reconstruction depends also on the part of the pavement not-resurfaced. For example, it is possible that, after reconstruction, a subgrade weaker than that set out in contract causes a lower expected life of the surfacing, or that a base stronger then that set out in contract causes a greater one. Given the above facts, it is possible to demonstrate the following (see table 1):

$$PA \cong C_S \cdot (R^{DS} - R^{ES}) + \delta_{AC} \cdot C_S \cdot [-(R^{ES+DS})] + \delta_{AD} \cdot C_S \cdot [R^{2DS}] + (C_S) \cdot \{[(R^D))/(1-R^{ODP})] - [(R^{EB}))/(1-R^{OCP})]\} + (C_B) \cdot \{[(R^D))/(1-R^{ODP})] - [(R^{EB}))/(1-R^{OCP})]\} \cong$$

$$\cong C_S \cdot (R^{DS} - R^{ES}) + (C_S) \cdot \{[(R^D))/(1-R^{ODP})] - [(R^{EB}))/(1-R^{OCP})]\} + (C_B) \cdot \{[(R^D))/(1-R^{ODP})] - [(R^{EB}))/(1-R^{OCP})]\} \quad (12)$$

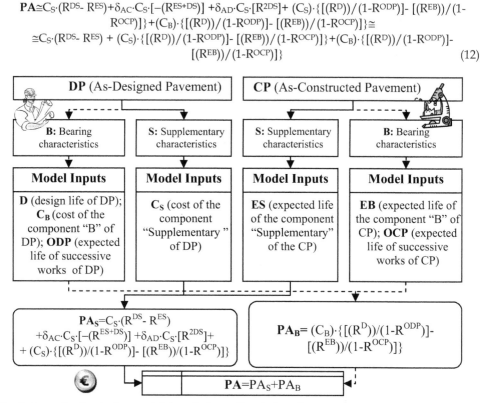

Fig. 3. A synthesis of the model (symbols are listed above).

The term containing C_B is the PA_B, while PA_S is given by the sum of the three terms containing C_S. Note that, when $D_S = E_S$, $E_B = E$, $C_S + C_B = C$, $O_{DP} = O_{CP} = O$ ($\delta_{AD} = \delta_{AD} = 0$), it follows (as in Burati et al, 2003; Weeds, 2001):

$$PA = (C) \cdot [(R^D - R^E))/(1-R^O)] \quad (13)$$

Table 2 illustrates expenditure flows. In order to test the validity of the hypotheses and the effectiveness of the method the following focal points can be focused:
i) In the case of an as-constructed pavement capable of withstanding the design loading, there is only a pay adjustment for non-conformities of supplementary characteristics:

$$if \ D = E_B \ and \ O_{DP} = O_{CP}, \ it \ is \ PA_B = 0 \quad (14)$$

or if

$$E_S = 0, \ E_B = D \rightarrow \infty, \ O_{DP} = O_{CP}, \ PA \rightarrow -C_S; \quad (15)$$

ii) The model is conceived for an expected life of the supplementary component less than that of the bearing component (as-constructed pavement); this limitation is quite obvious and agrees with the state of the art of pavement constructions. The equations system *must* contain also this equation:

$$E_S \leq E_B; \tag{16}$$

iii) As a consequence of the previous point, when the expected life of resurfacing/reconstruction is equal for both the as-Designed and the as-Constructed Pavement, if the expected life of the S component is greater than that of the B component, then the pay adjustment can be computed by the above-mentioned Ref.Eq. after (Weed, 2001):

$$if\ O_{DP}=O_{CP}\ and\ E_S>E_B,\ it\ is\ PA=C_{DP}\cdot(R^D-R^E)\cdot(1-R^O)^{-1} \tag{17}$$

iv) The pay adjustment must be compatible with the cost of the layers to resurface; from this another equation to be included in the equation system is derived. This equation originates from an intrinsic limitation of the model (in common with the previous model (Weed, 2001]); for example for DB=20, E_B=0, O_{DP}=O_{CP}=10, ES=DS, it is PA$\cong$-1.7$\cdot C_{DP}$ (which is inconsistent); the new model agrees with common sense if, when E_B=0, one puts also DB$\to\infty$ and O$\to\infty$, or/and with D=O. This supplementary equation is:

$$PA \geq -C_{DP} \tag{18}$$

vi) PA is substantially dependent on D-E (with E=EB=ES and O=OCP=ODP); it may be approximated, for example, by linear or quadratic relations:

$$PA\cdot(1-R^O)\cdot C_{DP}^{-1}= -0.02414284\cdot (D-E) \tag{19}$$

$$PA\cdot(1-R^O)\cdot C_{DP}^{-1}= -0.00049374\cdot(D-E)^2 - 0.0164080\cdot(D-E) - 0.00202513 \tag{20}$$

vii) PA is substantially dependent on OCP-ODP; for D=20, with E=EB=ES=15, it may be approximated, for example, by this linear relation:

$$PA\cdot C_{DP}^{-1}= -0.1601\cdot (OCP-ODP)-0.3753 \tag{21}$$

viii) PA% is affected by the difference (R^D-R^{ES}); this relationship depends on the thickness of resurfacing/reconstruction; if f_3=C_3/C_1, with f=C_2/C_1, OCP=ODP=O, D=EB, and n is the number of layers resurfaced:

$$PA\%= (1-fh)\cdot(1+f+f_3)^{-1}\cdot(R^D-R^{ES}),\ for\ n=3 \tag{22}$$

$$PA\%= (1-fh)\cdot(1+f)^{-1}\cdot(R^D-R^{ES}),\ for\ n=2 \tag{23}$$

$$PA\%= (1-fh)\cdot(R^D-R^{ES}),\ for\ n=1,\ where\ (1-fh)>(1-fh)\cdot(1+f)^{-1}>(1-fh)\cdot(1+f+f_3)^{-1} \tag{24}$$

3.3 Life expectancy of surface properties

This section deals with the estimate of the life expectancy for the different surface properties (see Figs 4-8, Tabs 2 and 3, equations 25-36). The author is aware that, if inadequately used, even the new model could cause misevaluations(Praticò, 2007). Some tools to optimize the estimate process are proposed below. The estimate of ES (as a minimum life expectancy for the various supplementary characteristics) can be obtained from quality control/assurance tests, if the time-dependence of the supplementary characteristics, for given traffic, is predictable. There are many effects and related indicators that can be considered eligible as supplementary characteristics (AA.VV., 2005): drainagebility, friction (polishing), noise, texture, splash & spray, raveling, reflectivity, chemical spill tolerance, faulting (difference in

AD, As-designed pavement		AC, As-constructed pavement		AD vs. AC
Times	Expenditure	Times	Expenditure	Pay Adjustment
DS	$C_S \cdot (R^{DS})$	ES	$C_S \cdot (R^{ES})$	$C_S \cdot (R^{DS}) - C_S \cdot (R^{ES})$
$2DS$	$\delta_{AD} \cdot C_S \cdot R^{2DS}$	$ES+DS$	$\delta_{AC} \cdot C_S \cdot [(R^{ES+DS})]$	$\delta \cdot C_S \cdot [(R^{ES+DS}) - R^{2DS}]$
DB	$(C_B + C_S) \cdot (R^{DB})$	EB	$(C_B + C_S) \cdot (R^{EB})$	$(C_B + C_S) \cdot (R^{DB} - R^{EB})$
$DB + O_{DP}$	$(C_B + C_S) \cdot (R^{DB+ODP})$	$EB + O_{CP}$	$(C_B + C_S) \cdot (R^{EB+OCP})$	$(C_B + C_S) \cdot (R^{DB+ODP} - R^{EB+OCP})$
$DB + 2 \cdot O_{DP}$	$(C_B + C_S) \cdot (R^{DB+2 \cdot ODP})$	$EB + 2 \cdot O_{CP}$	$(C_B + C_S) \cdot (R^{EB+2OCP})$	$(C_B + C_S) \cdot (R^{DB+2 \cdot ODP} - R^{EB+2OCP})$
$DB + n \cdot O_{DP}$	$(C_B + C_S) \cdot (R^{DB+nODP})$	$EB + n \cdot O_{CP}$	$(C_B + C_S) \cdot (R^{EB+nOCP})$	$(C_B + C_S) \cdot (R^{DB+nODP} - R^{EB+nOCP})$
Sum	$C_S \cdot (R^{DS})$ $+ (C_B + C_S) \cdot (R^{DB})/(1 - R^{ODP})$		$C_S \cdot (R^{ES})$ $+ (C_B + C_S) \cdot (R^{EB})/(1 - R^{OCP})$	$C_S \cdot (R^{DS} - R^{ES})$ $+ \delta_{AC} \cdot C_S \cdot [-(R^{ES+DS})] +$ $\delta_{AD} \cdot C_S \cdot [R^{2DS}] +$ $(C_B + C_S) \cdot \{[(R^{DB}))/(1 - R^{ODP})] - [(R^{EB}))/(1 - R^{OCP})]\}$

Table 1. A synopsis of times and expenditures flows for AD and AC.

elevation across a joint), pitting, resistance to wearing, etc. Let Esi be the expected life in relation to the i-th characteristic. Under these hypotheses, Es will be the minimum value among the Esi. In the light of the above facts, if i=1 (for example friction) and the as-designed target is q_a, while the as-constructed value of friction is $q_b < q_a$, in the case of linear law over the time it results: a) expected life of the as-designed friction course: $ES_a = (T - q_a)/m$, where T stands for minimum level and m is the gradient (negative); b) expected life of the as-constructed friction course: $ES_b = (T - q_b)/m$, where T stands for minimum level; c) loss of expected life: $ES_a - ES_b = (q_b - q_a)/m$. For example, if m=-2, T=35, q_a=55, q_b=45, it results $ES_a - ES_b$ =5 years.

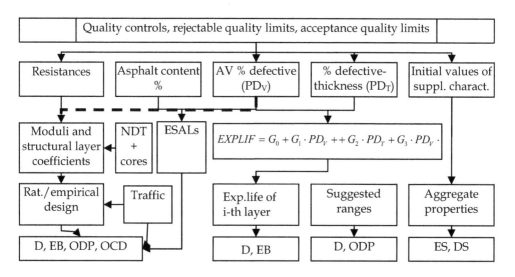

Fig. 4. Expected life: tools to optimize the estimate (symbols are listed in table above).

As is well known, skid resistance changes over time. Typically it increases in the first two years following construction as the roadway is worn away by traffic and rough aggregate surfaces become exposed, then decreases over the remaining pavement life as aggregates become more polished. Skid resistance is also typically higher in the fall and winter and lower in the spring and summer. This seasonal variation is quite significant and can severely skew skid resistance data if not compensated for. Skid resistance deterioration rate depends on the combination of many factors (aggregate properties and gradation, alignment, detritus build-up, rainfall and traffic characteristics, etc) and this can affect substantially any attempt at prediction and modelling. On the basis of the international literature on this topic the following equation can be here derived for skid resistance (F) drop due to pavement wear:

$$F = F_1 + F_2 \cdot e^{\left[\frac{Y}{Y} \cdot Ln(0.001)\right]} + F_3 \cdot \prod_{\tau=1}^{\tau\tau} sen(2 \cdot \pi \cdot fre_i - \phi_i) \tag{25}$$

where i) Seasonal, monthly, daily effects and variations are considered by the different frequencies (fre_i) (Diringer and Barros, 1990; Wilson and Kirk, 2005); initial increase is not explicitly considered here; ii) The terminal value F_1 depends on aggregate shore hardness (both average value and coefficient of variation); this value seems to depend on Polishing Stone Value (PSV, British Standard BS 812, Italian standard CNR BU 140/92) and on Los Angeles (or micro-Deval) (Dupont and Turenq, 1993); both the Los Angeles abrasion number and the PSV, alone, do not correlate well with field performance (Dupont and Turenq, 1993); iii) time in years to reach a quasi-constant friction (Y*) can correspond to 2 million cumulative vehicle passes (Diringer and Barros, 1990) and is usually two to four years, depending on traffic and aggregate properties; iv) $F_2 \cdot F_1^{-1}$ can be estimated in 0.8~1.1 for sedimentary rocks and 0.5~1.4 for igneous rocks (data referred to BPN); v) $F_3 \cdot F_1^{-1}$ can be estimated equal to 0.1~0.3 (Diringer and Barros, 1990). Figure 5 (left) provides a simplified depiction of these hypotheses on friction time-dependence, where Y*=4, F_1=47; F_2=38, fre_1=1year^{-1}, φ_1=φ_2=0; fre_2=4 year^{-1}, $F_3 \cdot F_1^{-1}$=0.22; dotted curve refers to fre_2=0. Figure 5 (right) shows time-dependence for two friction indicators (initial increase is not represented (Brosseaud and Roche, 1997; Kokkalis and Panagouli, 1998). A possible relationship between SFC (Side Friction Coefficient, range 0-1) and SN (Skid Number, range 0-1) (Ullidtz, 1987) is as follows (see figure 5):

$$SFC=-0.014+1.516 \, SN \tag{26}$$

The minimum value of SFC over the time can be also estimated through the following equation (see figure 6):

$$Min \, SFC=0.024-0.663*10^{-4} \, QCV+0.01PSV \tag{27}$$

where QCV is the number of commercial vehicles/lane/day, PSV is the Polishing Stone Value (Ullidtz, 1987). For limestone, the decrease of SFC over the time as a function of N (number of heavy vehicle equivalents in millions) and SFCi (initial value of SFC) can be expressed as (see figure7):

$$\Delta SFC=-0.48N^{0.373}(SFCi-3) \tag{28}$$

while for basalt as (figure 7):

$$\Delta SFC = -0.30 N^{0.503}(SFC_i - 3) \tag{29}$$

(Flintsch et al, 2001) proposed the following model:

$$SN(64)_S = 26.865 + 2.079 \cdot Binder + 1.601 \cdot PP200 + 1.03 \cdot VTM \tag{30}$$

$$SN(64)_R = 104.211 - 4.356 \cdot NMS + 0.1833 \cdot VTM \tag{31}$$

where $SN(64)_S$ stands for Skid Number measured at 64Km/h for smooth tires, Binder stands for binder code (-1 for PG 64-22, 0 for PG 70-22, 1 for PG 76-22), PP200 is the percentage of material passing the #200 sieve, VTM represents the total voids in the mix, $SN(64)_R$ stands for Skid Number measured at 64Km/h for ribbed (R) tires, NMS is the Nominal Maximum Size. Note that the model after Flintsch et alia refers to the starting point of diagrams (as-constructed value). In particular, for Binder=-1, NMS=12.5, PP200=5, VTM=20 it results $SN64_S$= $SN64_R$=53 and for Binder=-1, NMS=19, PP200=5, VTM=6, it results $SN64_S$= 39 and $SN64_R$=23. As for texture, (Flintsch et al, 2001) proposed the following model:

$$MPD = -2.896 + 0.2993 NMS + 0.0698 VMA \tag{32}$$

where MPD stands for Mean Profile Depth, NMS for Nominal Maximum Size and VMA for Voids in Mineral Aggregates. Note that in this case the law doesn't provide the variation over the time. Another model for texture depth was developed (Arnold et al, 2005) according to the following algorithm:

$$MPD = k1 - k2 \cdot \log(N), \tag{33}$$

where $k1$ and $k2$ are constants and N is number of wheel passes. The constants k_1 and k_2 in the equation (the Patrick equation) have been calculated for two different cases (see figure 7). The Surface wear due to the combined action of salt and traffic can be estimated through the following model (Ullidtz, 1987, Praticò et al, 2010):

$$RDW = 2.48 \cdot 10^{-5} \cdot PASS^{1.02} \cdot CW^{-0.46} \cdot S^{1.22} \cdot SALT^{0.32} \tag{34}$$

Where RDW is the rut depth due to studded tires in mm, PASS is the number of vehicles with studded tires in one direction expressed in thousand, CW is the carriageway width in m, S is the vehicle speed in Km/h, and SALT is a variable for salting (2 salted, 1 unsalted, see figure). Based on other authors (Smith, 1979), with studded tires in the range 7-23%, concrete pavement wear can be considered as follows:

$$PW \approx 0.5 * Y \tag{35}$$

Where PW is the wear in mm, while Y stands for number of years. Also drainability values depend on the chosen indicator (Praticò and Moro, 2007a, 2008a). On the basis of the international literature on this topic, a typical curve for drainability is as follows (where I is drainability indicator, Y stands for years, α and χ are positive and β is negative; PEM indicates porous European mixes; EM: emergency lane):

$$I = \alpha \cdot e^{\beta \cdot Y} + \chi \tag{36}$$

It is important to note that, although some of these estimates may be considered approximate, all the inputs may be "conditioned" by the same methodology and the PA formula is strongly dependent on the differences D-E and O_{CP}-O_{DP}. This can help to reduce the possible conflicts between contractor and buyer without using simple but empirical formulas and models not well-grounded in logic.

Pavement	α	β	χ	Indicator I	I Range	Y Range (years)
PEM	0.469	-0.778	0.862	Percolation speed (cm/s)	0.79~1.5	0~3
PEM	0.449	-2.435	0.795	Percolation speed (cm/s)	0.57~1.6	0~3
PEM	1.049	-0.778	0.248	Permeability (cm/s)	0.3~1.3	0~3.75
PEM - EL	1.119	-1.312	0.168	Permeability (cm/s)	0.15~1.3	0~3.75

Table 2. Time-dependence of drainability (example).

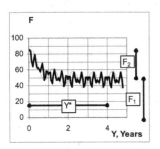

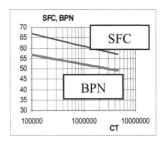

Note: F: Friction indicator; SFC: Side Friction Coefficient; BPN: British Pendulum Number

Fig. 5. Skid resistance (F=BPN) vs. time (Years) and skid resistance (SFC, BPN) versus equivalent Cumulative Traffic (CT).

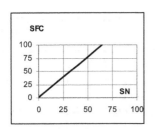

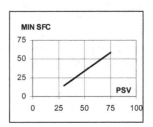

Fig. 6. Example of relationship SN vs. SFC and PSV vs. MINSFC.

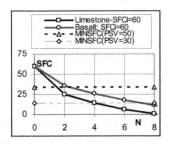

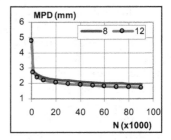

Fig. 7.Example of relationships N vs. SFC and MPD vs. number of axes.

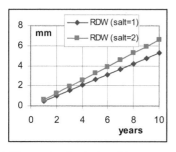 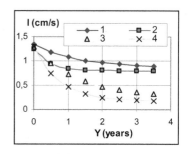

Fig. 8. Example of relationships years vs. wear (mm) or permeability.

As for noise mitigation, based on the international literature the following information is available:

Solution	E	IN	FINAL
DAC (dense asphalt concrete)	Variable	0	-2
PA (porous asphalt)	10-12	4	<3 (?)
TPA (two-layer porous asphalt)	9	6	4
SMA-like thin layers	9.5	4.7	3
Porous-type thin layers	8.5	5	3
E: Expected lifetime (years); IN: Initial noise reduction (db(A)); FINAL: Final/minimum noise reduction (db(A))			

Table 3. Duration of noise mitigation (SPB method).

3.4 Life expectancy of bulk properties
Both for unbound and HMA/PCC layers, volumetrics, mechanistic performance and related indicators are often correlated and vary over the time. Equations 37-38, Figs 9-13, Tables 4 and 5 summarize several practical relationships. In the M-E PDG, the HMA layer modulus is characterized using the dynamic modulus (or backcalulated modulus from FWD data). In more detail, the dynamic modulus at a given loading time and temperature is assumed to be the elastic modulus in the response computation. PCC materials need a static modulus of elasticity adjusted with time. For chemically stabilized materials the elastic modulus or the resilient modulus is needed (lime-stabilized typical value: 45000psi). For unbound materials the resilient modulus is needed (39000 psi: very good; 10000psi: very poor). In the AASHTO guide 1993 the structural layer coefficients and the resilient modulus (subgrade) are used. Design life (D) of the as-Designed Pavement (DP), can be estimated from contract specifications. D can also be estimated on the basis of the design report (in which it is usually specified). Empirical or rational design can be used after having estimated structural layer coefficients or moduli from resistance thresholds set out in the contract (Marshall Stability, MS, for surface course, base course and bituminous treated bases, California Bearing Ratio (CBR), for untreated bases or subbases, Unconfined Compressive Strength – 7 day break (UCS) for cement treated bases, (Huang, 2003), see figures 9 to 13); by knowing thickness and traffic loading it is therefore possible to estimate D. Importantly, on the basis of the international literature (Van Til et al, 1972; Gaspard, 2000; Sebesta, 2005) and current practice, the modulus of cement treated bases

which corresponds to a given laboratory resistance (Unconstrained Compressive Strength – 7-day break) can vary to a great extent in relation to many factors (time from base construction, traffic, subgrade drainage, shrinkage cracks, etc). Figure 9 shows the relationship between resistances and structural layer coefficients according to (Van Til et al, 1972). Note that in the following figures MS indicates Marshall stability, M the modulus, a the structural layer coefficient, UCS is the unconstrained compressive strength- 7 day break, CBR is the california bearing ratio; ESAL indicates equivalent single axle load, AC stands for as-constructed. In principle, E_B, expected life of the bearing component of the as-Constructed Pavement, can be estimated as D (except where D is derived from thresholds in contracts or from design report). Moreover, it must be noted that for the asphalt layers there are empirical formulas that can be used to estimate D and E_B in function of Percent Defective of air voids and thickness (PD_V, PD_T, (Burati et al, 2003)), or in function of the air voids and asphalt content of the as-Constructed Pavement. Both for E_B and D, reliable information can be obtained from Non-Destructive Tests, NDT (e.g. Falling Weight Deflectometer), or /and laboratory tests on cores (resilient moduli – test methods AASHTO TP9-94-1B, ASTM D 4123, LTPP P07) (Giannattasio and Pignataro, 1983; Ullidtz, 1987). The estimate of ODP and OCP can be approached by the same methodologies above-mentioned for D and E_B respectively (see also tables 4 and 5). All these values can be modified (and costs upgraded) if extended service life gains for preventive maintenance treatment are considered (overband crack filling, crack sealing, single or double chip seal, slurry seal, microsurfacing, ultrathin, hot-mix asphalt overlay, hot-mix asphalt mill and overlay, etc.) and in relation to particular design philosophies.

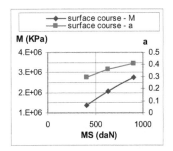
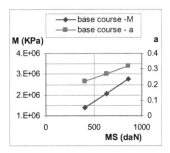

Fig. 9. Example of MS-M-a relationships, for surface courses and base courses.

Mr (psi)	a	a	a	UCS, Mpa	UCS,psi	MD, Mpa	CBR	MD, Mpa	CBR	MD, Mpa	CBR	MD, Mpa
all	CTB (1)	UB (2)	SB (3)	CTB (4)	CTB	CTB (5)	UB (6)	UB (7)	SB	SB (8)	SUB (9)	SUB (10)
9,128			0.06						6	31	6	29
12,628			0.09						17	**80**	8	40
15,000			0.11						30	144	10	48
15,750		0.07	0.11				10	47	40	192	11	**50**
18,300		0.08	0.13				17	**80**	64	309	12	59
20,548		0.10	0.14				25	122	100	479	14	66
25,000		0.12					52	248			**17**	80
30,000		0.14					100	480			20	96
33,582		0.15					150	722			22	107
39,000											26	125
45,000											30	144
599,081	0.15			2.5	362	**150**						
752,581	0.20			4.7	676	381						
864,318	0.23			6.0	867	521						

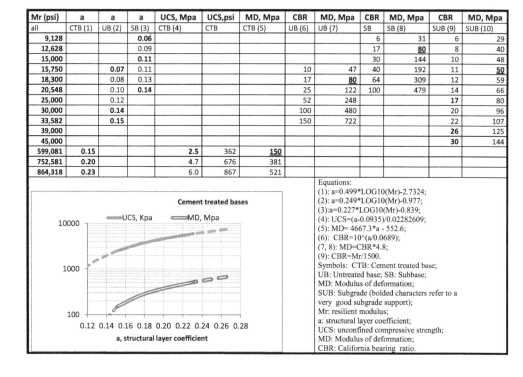

Equations:
(1): $a = 0.499 \cdot \mathrm{LOG10}(Mr) - 2.7324$;
(2): $a = 0.249 \cdot \mathrm{LOG10}(Mr) - 0.977$;
(3): $a = 0.227 \cdot \mathrm{LOG10}(Mr) - 0.839$;
(4): $UCS = (a - 0.0935)/0.02282609$;
(5): $MD = 4667.3 \cdot a - 552.6$;
(6): $CBR = 10^{(a/0.0689)}$;
(7, 8): $MD = CBR \cdot 4.8$;
(9): $CBR = Mr/1500$.
Symbols: CTB: Cement treated base;
UB: Untreated base; SB: Subbase;
MD: Modulus of deformation;
SUB: Subgrade (bolded characters refer to a
very good subgrade support);
Mr: resilient modulus;
a: structural layer coefficient;
UCS: unconfined compressive strength;
MD: Modulus of deformation;
CBR: California bearing ratio.

Fig. 10. Cement treated base courses, untreated bases, subbases, subgrades.

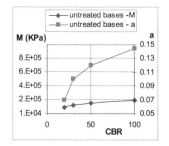
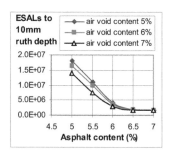

Fig. 11. Left (untreated bases): CBR vs. M and a. Right (surface course - fine gradation): ESALs to 10mm Ruth Depth vs. asphalt content (Epps et al. , 1999).

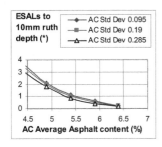

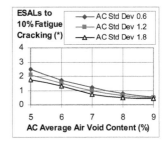

Fig. 12. Left: ESALs to 10mm vs. asphalt content. Right: ESALs to 10% vs. air void content (Surface courses (Epps et al. , 1999)) (*) expressed as multiple of target ESALs.

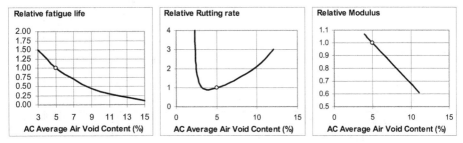

Fig. 13. Air Voids vs. Relative Fatigue Life (left) or Relative Rutting Rate (center) or . Relative Modulus (right) - Surface courses (Austroroads, 1999).

years to 1st rehabilitation ($\cong$D)	years to 2nd rehabilitation ($\cong$ODP)	years to 3rd rehabilitation ($\cong$ODP)	years to 4th rehabilitation ($\cong$ODP)
12~16 (base layer undrained) 15~20 (base layer drained)	10~12 (mill and overlay)	12~16 (full depth mill and overlay) 10~12 (mill and overlay)	10~12 (mill and overlay)

Table 4. Estimates of years to the n-th rehabilitation.

Note that the application of the PA formula depends on the ability to split the surface course into two parts. A tentative method to estimate t_{1B}, C_{1B}, C_B and C_S is to identify the component B of DP in a pavement with a design life D, but with a different friction course (this time with negligible surface properties, for example just the binder course). In particular, two tentative values of t_{1B} may be obtained by the following expressions:

$$t_{1B} \cdot M_{1B}^{1/3} \cong t_1 \cdot M_1^{1/3} \; ; \quad t_{1B} \cdot a_{1B} \cong t_1 \cdot a_1 \qquad (37)$$

where the modulus M_{1B} of the first layer of the B component of the as-Designed Pavement can be tentatively identified in M_2, the structural layer coefficient a_{1B} can be considered equal to a_2, and the thicknesses t_1 and t_2 are known. As above-mentioned, M_1, M_2 (moduli of the 1st and 2nd layer of the DP) and a_1, a_2 (structural layer coefficients of the 1st and 2nd layer of the DP) may be estimated by using correlation charts and algorithms in literature

references (Van Til et al, 1972; Huang, 2003) in function of the traditional resistance tests. Similarly, C_{1B}, C_B and C_S (€/m²), for a given t_{1B}, can be easily estimated by comparing the costs of the 1st and 2nd layer of the DP, given that the cost of the unit of volume of the layers 1B and 2 of the DP are the same. For example, if C_2 and t_2 are the cost and thickness of the second layer of the as-Designed Pavement respectively, then for h=0.5, f=0.6:
if

$$t_{1B}=h \cdot t_2 , \ C_2=f \cdot C_1, \ C_{1B} \cdot t_{1B}^{-1}=C_2 \cdot t_2^{-1} \ (€/m^3) \Rightarrow C_{1B}= f \cdot h \ \cdot C_1, \ C_S=C_1 \cdot (1- f \cdot h)$$

$$C_S=0.7 \cdot C_1, \quad C_{1B}=0.3 \ C_1 \quad C_B=0.3 \cdot C_1+C_2+...+C_n \quad\quad (38)$$

Treatment	expected Life (Years, Y)		Cost (€/ m²)	Cost for year €/(m².Y)
	min	max		
Crack seals (PM)	2	3	1.7	0.66
Fog Seals (PM)	3	4	0.8	0.24
Slurry seal (PM)	4	9	1.5	0.23
Microsurfacing (PM)	5	14	2.3	0.24
Chip seals (PM)	4	6	1.3	0.26
Thin hot mix overlay (PM/REH)	2	10	2.8	0.47
HMA– Dense Graded (5cm) (PM/REH)- (ODP→)	5	15	5.3	0.53
Asph. Rubber Hot Mix – Gap Graded (4-5cm) (PM/REH)-(ODP→)	10	20	5.9	0.40
37.5mm mill+37.5mm overlay (PM/REH) - (ODP→)	8	12	7.2	0.72
Milling (37.5mm) +100mm overlay (REH) - (ODP→)	18	22	9.5	0.48
Milling+thick overlay (75-180mm) (REH) - (ODP→)	18	25	10.9	0.51
Note. PM: Preventive Maintenance treatment; REH: REHabilitation				
After (Moulthrop et al, 1998; Hicks and Epps, 2005; Shober and Friedrichs, 2002)				

Table 5. Summary of a variety of maintenance and rehabilitation life expectancy and costs.

3.5 Experimental application
An experimental application was performed on a motorway in Southern Italy. Traditional quality characteristics of the mixes, for single lane and given layer, have been organized in function of the progressive abscissa. Many experimental devices were used in order to measure the actual characteristics of the as-constructed pavement. From the comparison between the requirements of the as-designed pavement (DP) and the actual characteristics of the as-constructed pavement (CP) the moduli for DP(@0) and CP(@0) were derived (see table 7). The ratio t_{1B}/t_2 was about 0.5; asphalt concrete thickness was 27 cm for DP and about 27-5+3.5=25.5cm for the B component of the pavement. Results were obtained by Kenpave™ [Huang, 2003] and successfully compared with the AASHTO Guide 1993 equation. By analyzing the drainability and friction data, $E_S=3$ was estimated, caused by insufficient drainability. Pay adjustments (in absolute and in percentage, as referred to the cost C_{DP}) are summarized in table 8.

	DP (@0)	DP, B(@0)	CP(@0)	DP(@E_B)	CP(@E_B)
Asphalt (27 or 25.5* cm)	$2.29 \cdot 10^6$	$2.65 \cdot 10^6$ (*)	$1.00 \cdot 10^6$	$2.29 \cdot 10^6$	$2.29 \cdot 10^6$
Cement treated (20 cm)	$1.00 \cdot 10^6$	$1.00 \cdot 10^6$	$1.30 \cdot 10^6$	$0.50 \cdot 10^6$	$0.70 \cdot 10^6$
Subbase + sub-grade	$0.10 \cdot 10^6$	$0.10 \cdot 10^6$	$0.13 \cdot 10^6$	$0.10 \cdot 10^6$	$0.11 \cdot 10^6$
Life (years)	D=22	D=22	E_B=18	O_{DP}=12	O_{CP}=14
Output	D	t_{1B}/t_2=0.50	E_B	O_{DP}	O_{CP}

Table 7. Case-history (years are rounded to the nearest integer; moduli in KPa).

Input	INT	0.08	C_2, asphalt, €/ m²	4,25
	INF	0.04	C_3, asphalt, €/ m²	7,27
	R	0.963	Cost of the cement treated, €/ m²	6,92
	D (years)	22	Cost of the granular sub-base, €/ m²	3,83
	E_B (years)	18	Overall Cost €/m²	29,09
	E_S (years)	3	Asphalt cost €/ m²	18,34
	O_{DP} (years)	12	C_B, €/ m²	13.65
	O_{CP} (years)	14	C_S, €/ m²	4.69
	t_{1B}/t_2	0,5	C_B%	74.42
	C_1, asphalt, €/m²	6,82	C_S%	25.58
Output	PA (€/m²)	-2.51		
	PA_B (€/m²)	-0.52		
	PA_S (€/m²)	-1.99		
	PA%	-13.70		
	PA_B %	-2.85		
	PA_S %	-10.85		

Table 8. PA determination –Inputs and outputs *(the percentages are referred to C_{DP}).*

4. Main findings

In the light of above facts, the main findings of the study can be summarized in terms of drawbacks and point of strengths. Air void content has a vital role in QC/QA. A decrease in porosity (or effective porosity) yields in-lab specific gravities that converge toward the maximum theoretical specific gravity G_{mm} (convergence of in-lab measurements). In contrast, gravities determined through non-nuclear portable devices often increase due to the high dielectric constant of water (divergence of in-site measures).

Uncertainties in the determination of the expected life of the different characteristics and complexity are the main drawbacks of the models of PA based on LCCA. On the contrary, though more research is needed, analyses proved that the LCCA-based models are able to give a solution well-grounded in logic, even in cases where supplementary characteristics, such as drainability or friction, may have a premature failure. In particular, when only the bearing characteristics of asphalt layers are considered, analyses proved that the complex model (in which also supplementary characteristics are considered) gives the same solution as an existing model (Weed, 2001). The effectiveness and importance of the complex models increases when thin resurfacing/reconstruction is considered. The conceptual framework of the complex models, finally, has the potential for considering other supplementary characteristics such as for example the albedo. Analyses and experimental applications proved that the complex models quantify the dependence of the pay adjustment on all the layers. Another point of interest is that life expectancy differences govern PA estimates: this can play a key role in reducing possible conflicts between buyer and contractor. Future research will aim to consider the consequences of the new model here set out in terms of relationship between probability of acceptance and expected lives.

5. References

Alvarez AE, Epps Martin A, Estakhri C. Internal structure of compacted permeable friction course mixtures. Construction and Building Materials, 24 (2010) 1027–1035.

Arnold G., Steven B., Alabaster D. & Fussell A. (2005). Effect on Pavement Wear of Increased Mass Limits for Heavy Vehicles – Stage 4 - Land Transport New Zealand Research Report 280.

Austroroads and Australian Asphalt pavement Association, Air voids in asphalt, 1999.

Boscaino, G., Pratico, F.G. , A classification of surface texture indices of pavement surfaces | [Classification et inventaire des indicateurs de la texture superficielle des revêtements des chaussées], 2001, Bulletin des Laboratoires des Ponts et Chaussees (234), pp. 17-34+123+125+127.

Brosseaud Y., Roche J-P (1997). Chantier experimental d'enrobés drainants, Bilan après huit annèes de traffic sur l'A63, Revue Générale des routes et des aerodromes, n.755, pp.47-52.

Brown ER. Hainin MR, Cooley A, Hurley G. Relationship of Air Voids, Lift Thickness, and Permeability in Hot Mix Asphalt Pavements. NCHRP Report 531, National Center for Asphalt Technology — Auburn University, Auburn, AL, Transportation Research Board, Washington, D.C. 2004.

Burati J., L., Weed R. M., Hughes C. S., Hill H. S.(2003). Optimal procedures for quality assurance specifications, Final Report n. FHWA-RD-02-95, Office of research, development, and technology, Federal Highway administration (FHWA - COTR: Peter A. Kopac) – 6300 Georgetown Pike, McLean, VA 22101-2296.

Burati, J., L., Jr (2005). Risks with Multiple Pay Factor Acceptance Plans, TRB 2005 Annual Meeting, January.

Cooley LAJr, Prowell BD, Hainin MR, Buchanan MS, Harrington, J. Bulk specific gravity round-robin using the corelok vacuum sealing device. National Center for Asphalt Technology Report 02-11, FHWA-IF-02-044, November 2002.

Crouch LK, Badoe DA, Cates M, Borden TA, Copeland AR, Walker CT, Dunn T, Maxwell RA, Goodwin WA. Bulk specific gravity of compacted Bituminous mixtures:

finding a more widely applicable method. Final Report, State of Tennessee, Department of Transportation, July 2003.

Deacon John A., Monismith Carl L., and Harvey John T. (1997). California department of transportation, "Pay factors for asphalt-concrete construction: effect of construction quality on agency costs".

Di Benedetto H. et alia (1996). Mesures in situ et en laboratoire des écoulement dans les BBDr, modélisation et validation, Bulletin des Laboratoires des Ponts et Chaussées, 204.

Diringer K. T., Barros R. T. (1990). Predicting the skid resistance of bituminous pavements through accelerated laboratory testing of aggregates, Surface characteristics of roadways: International research and technologies, ASTM STP 1031, W.E. Meyer and J. Reichert, Eds, ASTM, Philadelphia.

Domenichini L., Fracassa A., La Torre F., Loprencipe G., Ranzo A., Scalamandrè A. (1999). Relationship between Road Surface Characteristics and Noise Emission, 1° Int.Coll. on Vehicle Tyre Road Interaction, Roma.

Dupont P., Tourenq C. (1993). Granulats et microrugosité, Bull. Liaison Labo. P. et Ch., Ref. 3719, mai-juin.

Epps J. A. et al. (1999). "The road to performance-related specifications", International Conference on Accelerated Pavement Testing, Reno, NV, October 18-20.

Flintsch G, Al-Qadi I.L., Davis R., McGhee K.K. (2001). Effect of HMA properties on pavement surface characteristics, 2002 FWD Users Group Presentations, Nevada.

Friedrichs D. A., Pavement design and programmino options: european design to maintain only, Wisconsin Department of Transportation Library, USA, 2002.

Gamache RW. FEA guides the way for sensor redesign. Machine Design, April 15, 2004. p. 25.

Gaspard K.J., Evaluation of Cement Treated Base Courses, Technical assistance report Number 001TA, Louisiana transportation research center, Louisiana department of transportation and development, December, 2000.

Giannattasio P., Pignataro P., La caratterizzazione dei conglomerati bituminosi, dati di ingresso per il calcolo delle pavimentazioni, Scuola di specializzazione in infrastrutture aeronautiche, Napoli, 1983.

Huang Y.H. (2003). Pavement Analysis and Design, Pearson Prentice Hall, Upper Saddle River, NJ, USA.

Hughes C.S. et alia (2005). NCHRP Synthesis 346, State Construction Quality Assurance programs, Transportation Research Board, Washington D.C..

Khaled Ksaibati, Ph.D., P.E., Nathan E. Butts (2003). Asphalt pavement quality control/quality assurance programs in the United States, TRB 2003 Annual Meeting .

Kokkalis A. G., Panagouli O. K.(1998). Fractal evaluation of pavement skid resistance variations. II: surface wear, Chaos, solitons & fractals, Vol.9, 11, pp.1891,1889, Pergamon Press.

Kvasnak AN, Williams C, Ceylan H, Gopalakrishnan K. Investigation of electromagnetic gauges for determining in-place HMA density. IHRB Project TR-547, CTRE Project

05-233, Final Report, Iowa State University, Ames; Iowa Department of Transportation, 2007.

Leahy et al. Committee on Management of Quality Assurance (A2F03) (2002). Glossary of Highway Quality Assurance terms, Transportation Research Circular number E-C037, TRB National research Board, 2101 Constitution Avenue, NW, Washington, DC 20418, April.

Main italian laws and standards concerning pay adjustments (1998-2000). a) L.11.02.94, n. 109; b) L. 18 novembre 1998, n. 415 - c.d. L. Merloni ter; c) Capitolato Speciale d'appalto ANAS 1998 e succ.– artt.12, 13, 19; d)D.P.R. 21.12.99 n. 554 (Reg. di attuazione); e) Norme tecniche di tipo prestazionale per capitolati speciali d'appalto (CIRS, Centro sperimentale Interuniversitario di Ricerca Stradale), 2000; Capitolato prestazionale ANAS 2008; Capitolato Autostrade.

Megali, G., Cacciola, M., Ammendola, R., Moro, A., Praticò, F.G., Morabito, F.C., Assessing reliability and potentiality of nonnuclear portable devices for asphalt mixture density measurement, 2010, Journal of Materials in Civil Engineering 22 (9), pp. 874-886.

Mohammad LN, Herath A, Wu Zhong, Cooper SA. Comparative study of factors influencing the permeability of hot-mix asphalt mixtures. Volume 74E Electronic Journal 2005, AAPT Publications, 2005.

Montepara A, Virgili A. La determinazione della massa volumica e dei vuoti nei conglomerati bituminosi aperti. Atti Convegno S.I.I.V. Ancona, 1996.

Moulthrop J.S., Hicks G.R., Delaiden J.L., Selecting a flexible pavement preventive maintenance, 1998 Western Pavement maintenance Forum, Sacramento, CA, USA, January, 1998.

Muench, S.T. and Mahoney, J.P.(2001). A Quantification and Evaluation of WSDOT's Hot Mix Asphalt Concrete Statistical Acceptance Specification. WA-RD 517.1. Washington State Department of Transportation, Transportation Center (TRAC). Seattle, WA.

Niagara Peninsula, "Special provisions - contract items", 2004.

Praticò F.G., How should payment adjustments be assessed when both surface and mechanical defects are involved? a synergetic study on theory and experiments, SURF 08 – 6th Symposium on surface characteristics, Portoroz, Slovenia, 20/22 October -2008.

Praticò F.G., Quality and timeliness in highway construction contracts: a new acceptance model based on both mechanical and surface performance of flexible pavements, Construction Management and Economics, Volume 25, Issue 3, pages 305 – 313, Routledge - Taylor and Francis 2007, March.

Praticò FG, Moro A, Ammendola R. Modeling HMA Bulk Specific Gravities: A Theoretical and Experimental Investigation. International Journal of Pavement Research and Technology 2/3 (2009) 115-122.

Praticò, F.G., Moro, A , Measurement of air void content in hot mix asphalts: method and core diameter dependence, 2011b, Construction and Building Materials, doi: 10.1016/j.conbuildmat.2011.06.032.

Praticò FG, Moro A. Permeability and volumetrics of porous asphalt concrete: a theoretical and experimental investigation. International Journal: Road Materials and Pavements Design, Vol. 8 N 4/2007a.

Praticò, F.G., Ammendola, R., Moro, A., Factors affecting the environmental impact of pavement wear, 2010, Transportation Research Part D: Transport and Environment 15 (3), pp. 127-133

Praticò, F.G., Casciano, A., Tramontana, D., Pavement life-cycle cost and asphalt binder quality: Theoretical and experimental investigation, 2011a, Journal of Construction Engineering and Management 137 (2), pp. 99-107

Praticó, F.G., Moro, A., Ammendola, R., Factors affecting variance and bias of non-nuclear density gauges for porous european mixes and dense-graded friction courses, 2009a, Baltic Journal of Road and Bridge Engineering 4 (3), pp. 99-107+Ia+IIA+IIIA.

Praticò, F.G., Moro, A., Ammendola, R., Potential of fire extinguisher powder as a filler in bituminous mixes, 2010c, Journal of Hazardous Materials 173 (1-3), pp. 605-613

Praticò, F.G., Moro, A., Flow of water in rigid solids: Development and experimental validation of models for tests on asphalts, 2008a, Computers and Mathematics with Applications 55 (2), pp. 235-244.

Praticò, F.G., Moro, A., In-lab and on-site measurements of hot mix asphalt density: Convergence and divergence hypotheses, 2011, Construction and Building Materials 25 (2), pp. 1065-1071.

Praticò, F.G., Moro, A., Permeability and volumetrics of porous asphalt concrete: A theoretical and experimental investigation, 2007b, Road Materials and Pavement Design 8 (4), pp. 799-817

Praticò, F.G., Tramontana, D., Casciano, A., Attinà, A. , Geosynthetics and pavement life cycle: An analysis through the M-E PDG, 2010a, Geotechnical Special Publication (207 GSP), pp. 302-309.

Praticò, F.G., Tramontana, D., Casciano, A., Bitumen quality, pavement LCCA and contractor's expectations, 2010b, Geotechnical Special Publication (203 GSP), pp. 523-529

Rao C, Von Quintus HL, Schmitt R. Calibration of Non-nuclear Density Gauge Data for Accurate In-Place Density Prediction. Session 596. 86th TRB annual meeting, Washington D.C., January 2007.

Romero P. Evaluation of Non-Nuclear Gauges to Measure Density of Hot-Mix Asphalt Pavements. Pooled Fund Study, Final Report, The University of Utah Department of Civil and Environmental Engineering July 18, 2002.

Salt (1997). Research on skid resistance at the transport and road research laboratory (1972 to 1977), TRR 622, TRB, Washington, DC.

Sargand SM, Kim Sang-Soo, Farrington SP. Non-Nuclear Density Gauge Comparative Study. Draft Final Report. Ohio Research Institute for Transportation and the Environment, 114 Stocker Center, Ohio University, Athens, OH, 45701-2979, August 2005.

Sebesta S., Use of microcracking to reduce shrinkage cracking in cement treated bases, Transportation Research Board 2005 Annual Meeting, 2005.

Shober S.F., Friedrichs D.A. (2002). A Pavement Preservation Strategy, Wisconsin Department of ransportation Library, USA.

Siegwart, R. (2001). Indirect Manipulation of a Sphere on a Flat Disk Using Force Information. *International Journal of Advanced Robotic Systems,* Vol.6, No.4, (December 2009), pp. 12-16, ISSN 1729-8806

Smith R.D. (1979) Pavement wear and studded tire use in Iowa, Final Report, Iowa highway research Board, Project HR-148.

Spellerberg P, Savage D. An investigation of the cause of variation in HMA Bulk Specific Gravity test results using non-absorptive aggregates. National Cooperative Highway Research Program Web Document 66 (Project 9-26 (Phase 2), July 2004.

State of Florida Department of Transportation (2004). Asphalt concrete friction courses, December.

TransTech Systems Inc. Effect of Water and Temperature on Hot Mix Asphalt Density Measurement using Electromagnetic Sensing. TransTech Technical Note 0301, Schenectady, January 15, 2003. Available online:
http://www.transtechsys.com/products/pro_lib_pqi.htm.

Uddin, M., Mohboub, K.C., Goodrum, P.M. (2011). "Effects of Nonnormal Distributions on Highway Construction acceptance Pay Factor Calculations", Journal of Construction Engineering and Management, volume 137, n° 2, pp. 108-118.

Ullidtz, P. (1987). Pavement Analysis. Elsevier, Amsterdam.

Van Til, C.J., McCullough B.F., Vallerga B.A., Hicks R.G., Evaluation of AASHO Interim guides for of pavement structures, NCHRP 128, Highway Research Board, 1972.

Weed Richard M. (2001). "Derivation of equation for cost of premature pavement failure", 80th TRB annual meeting.

Weed Richard M., Kaz Tabrizi , Conceptual framework for pavement smoothness specification, TRB 2005 Annual Meeting, January, 2005.

Western Federal Lands Highway Division FP Specification Change (2004). "Pavement Smoothness/Roughness".

Whiteley Leanne, Susan Tighe, Zhanmin Zhang, Incorporating Variability into Pavement Performance, Life Cycle Cost Analysis and Specification Pay Factors, 84th Annual Meeting - January 9-13, 2005.

Williams SG. Non-Nuclear Methods for HMA Density Measurements. MBTC 2075, Final Report, University of Arkansas, June 2008.

Permissions

The contributors of this book come from diverse backgrounds, making this book a truly international effort. This book will bring forth new frontiers with its revolutionizing research information and detailed analysis of the nascent developments around the world.

We would like to thank Dr. Ahmed Badr Eldin, for lending his expertise to make the book truly unique. He has played a crucial role in the development of this book. Without his invaluable contribution this book wouldn't have been possible. He has made vital efforts to compile up to date information on the varied aspects of this subject to make this book a valuable addition to the collection of many professionals and students.

This book was conceptualized with the vision of imparting up-to-date information and advanced data in this field. To ensure the same, a matchless editorial board was set up. Every individual on the board went through rigorous rounds of assessment to prove their worth. After which they invested a large part of their time researching and compiling the most relevant data for our readers. Conferences and sessions were held from time to time between the editorial board and the contributing authors to present the data in the most comprehensible form. The editorial team has worked tirelessly to provide valuable and valid information to help people across the globe.

Every chapter published in this book has been scrutinized by our experts. Their significance has been extensively debated. The topics covered herein carry significant findings which will fuel the growth of the discipline. They may even be implemented as practical applications or may be referred to as a beginning point for another development. Chapters in this book were first published by InTech; hereby published with permission under the Creative Commons Attribution License or equivalent.

The editorial board has been involved in producing this book since its inception. They have spent rigorous hours researching and exploring the diverse topics which have resulted in the successful publishing of this book. They have passed on their knowledge of decades through this book. To expedite this challenging task, the publisher supported the team at every step. A small team of assistant editors was also appointed to further simplify the editing procedure and attain best results for the readers.

Our editorial team has been hand-picked from every corner of the world. Their multi-ethnicity adds dynamic inputs to the discussions which result in innovative outcomes. These outcomes are then further discussed with the researchers and contributors who give their valuable feedback and opinion regarding the same. The feedback is then collaborated with the researches and they are edited in a comprehensive manner to aid the understanding of the subject.

Apart from the editorial board, the designing team has also invested a significant amount of their time in understanding the subject and creating the most relevant covers. They scrutinized every image to scout for the most suitable representation of the subject and create an appropriate cover for the book.

The publishing team has been involved in this book since its early stages. They were actively engaged in every process, be it collecting the data, connecting with the contributors or procuring relevant information. The team has been an ardent support to the editorial, designing and production team. Their endless efforts to recruit the best for this project, has resulted in the accomplishment of this book. They are a veteran in the field of academics and their pool of knowledge is as vast as their experience in printing. Their expertise and guidance has proved useful at every step. Their uncompromising quality standards have made this book an exceptional effort. Their encouragement from time to time has been an inspiration for everyone.

The publisher and the editorial board hope that this book will prove to be a valuable piece of knowledge for researchers, students, practitioners and scholars across the globe.

List of Contributors

Ahmed Badr Eldin
Sigma Pharmaceutical Corp., Egypt

Isin Akyar
Acibadem University Faculty of Medicine Department of Medical Microbiology, Turkey

Mohamed K. Omar
Nottingham University Business school Malaysia, Malaysia

Sharmeeni Murugan
Faculty of Engineering & Technology Multimedia University, Malaysia

N. Phaswana-Mafuya
Human Sciences Research Council, Port Elizabeth, South Africa
Office of the Deputy Vice Chancellor: Research and Engagement, Nelson Mandela Metropolitan University, Port Elizabeth, South Africa

I. Senekal and S. Munyaka
University of Fort Hare, South Africa

A. S. Davids
Human Sciences Research Council, Port Elizabeth, South Africa

Jordy Coffa
MRC-Holland, The Netherlands
Free University Amsterdam, The Netherlands

Joost van den Berg
Berg IT solutions, The Netherlands

James B. Stribling
Tetra Tech, Inc., Center for Ecological Sciences, Owings Mills, Maryland, USA

Bor-Wen Tsai and Carl L. Monismith
University of California at Berkeley, USA

Jiangmiao Yu
University of California at Berkeley, USA
South China University of Technology, PRC

Catherine D. Kashork, Lisa G. Shaffer and Kyle S. Sundin
Signature Genomics Laboratories, USA

Yuehui Ma, Weijun Guan, Xiangchen Li, Xiaohong He, Yabin Pu, Qianjun Zhao, Dapeng Jin, Shen Wu, Taofeng Lu, Xiaohua Su and Chunyu Bai
Institute of Animal Sciences, Chinese Academy of Agricultural Sciences, Beijingm, PR China

Filippo G. Praticò
University Mediterranea at Reggio Calabria, Italy

Printed in the USA
CPSIA information can be obtained
at www.ICGtesting.com
JSHW011409221024
72173JS00003B/475

9 781632 400161